THE MICHIGAN CALCULUS PROGRAM

INSTRUCTOR TRAINING MATERIALS

for
Cooperative Learning
Homework Teams
Interactive Lecturing
Teaching Writing

•

•

•

Beverly Black
Pat Shure
Doug Shaw

Department of Mathematics
The University of Michigan

John Wiley & Sons, Inc.
New York • Chichester • Brisbane • Toronto • Singapore

ISBN 0-471-16521-2

Printed in the United States of America

10 9 8 7 6 5 4 3 2 1

Printed and bound by Odyssey Press, Inc.

These materials (including videos of classes in progress) are used at the University of Michigan to train instructors. Examples and lesson plans are geared to the text, Calculus by Hughes-Hallett, Gleason, et al.

Introduction

The Mathematics Department at the University of Michigan has made a commitment to making the introductory calculus course a vital part of our undergraduate students' education. We have changed both the content of the course and the ways we teach it by drawing ideas simultaneously from calculus reform nationwide and from the cooperative learning movement. Our text, with its fresh approach to calculus, gives us the opportunity to strengthen students' grasp of the underlying concepts. Students are taught to read the textbook and write full essay answers to many of the interesting problems. In our classes students work together actively, and outside of class they meet frequently to solve their team homework problems.

When we first engaged in this reform project, much of this was unfamiliar for both faculty and teaching assistants. So we set out to design a Professional Development Program which would help instructors learn the skills they needed to run a more student-focused course. Currently, any instructor who is new to the Michigan Calculus Program attends an intensive week of training before classes begin. After the term starts, we hold weekly follow-up workshops sequenced to complement what is happening as the course progresses. In addition, the training staff visits classrooms to provide each instructor with the necessary ongoing feedback.

This *Guide* gives a detailed description of each of the pre-term week of sessions including descriptions of all the videotapes and necessary materials. In addition, you will find outlines of some follow-up workshops to conduct during the term. At Michigan, the faculty and experienced graduate student teaching assistants who facilitate the Professional Development Program use the materials in the *Guide* as a place to begin. They then add their own personal touches and experiences to make the sessions come alive.

Most of the program is designed around a standard pattern. Through many years of experimenting, we have found that the best way to teach a skill is to describe it, demonstrate how to use it (or show a videotape of the skill being used in the classroom), discuss the demonstration, give the instructors an opportunity to practice the skill, and then give them feedback on their efforts. Throughout the program we try to keep the instructors actively involved and encourage them to exchange ideas about teaching and what it takes for students to learn. The program will be especially successful if several experienced instructors can share their successes, problems, and worries.

The Professional Development Program need not be unique to the University of Michigan. Any Mathematics Department which wants to use a reform text combined with active cooperative learning will find these materials easy to adapt. Most of the ideas and techniques can be useful for anyone who wants to reexamine the process of teaching.

While developing the program at Michigan, we relied on the support of the University administration which recognized that instructors would need considerable assistance in making such substantial changes. Also, the Center for Research on Learning and Teaching has cosponsored this project from the beginning. We are indebted to the National Science Foundation for partial funding of this project. Authors take full responsibility for all ideas contained in this handbook.

We are also grateful to the many math faculty members and graduate students who have gone through the training program and given us advice on how to improve it. We are especially indebted to Morton Brown who worked with us every step of the way, Doug Shaw who took the lead in developing the videotapes, Bob Megginson, Tony Giaquinto, and the many others who helped us design and facilitate the Professional Development Program.

Contents

(over)

Appendix VII	a. Group Observation Worksheet b. Two Sample Problems for Groups to Work
Appendix VIII	a. Overhead: "Organize the Information Flow" b. Overhead: "Plan Carefully and Use the Time Well" c. Sample Interactive Lesson Plans
Appendix IX	a. Handout: "How Doug Marks Papers" b. Handout: "A Few Suggestions on Grading"
Appendix X	a. Handout: "Why Get Students to Write" b. Overhead of Student's Work Without Written Explanations c. Overhead of Student's Work with Written Explanations

The University of Michigan
Mathematics Department Professional Development Program - Fall 1995

8:30	Monday August 28	Tuesday August 29 I	Tuesday August 29 II	Wednesday August 30	Thursday August 31	Friday September 1
9:00	Continental Breakfast	Computers, Library	Individual Practice Giving Explanations	Individual Practice in Asking and Answering Questions	Individual Practice in Using Cooperative Learning	Making Groups Work
9:30	Welcome Chair Overview of Week					
10:00	Course Philosophy Student Population	Break				
10:30		Individual Practice Giving Explanations	Break			
11:00	Distributions: Text, Calculators Materials		Computers, Library			Grades, Ethics, Values
11:30						
12:00	Lunch	Lunch		Lunch	Lunch	Lunch
12:30						
1:00	Creating an Atmosphere for Learning	Using Questions to Improve Thinking		Cooperative Learning in the Classroom	Homework Teams	Planning and Managing an Interactive Class
1:30						
2:00	Lecturing vs. Explaining					
2:30						Stump the Staff
3:00	Teaching with Technology	Grading System		Lecturing (how long and when)		Party
3:30						
4:00						
4:30						
5:00						

In collaboration with The Center for Research on Learning and Teaching, The University of Michigan

Professional Development Program

Mathematics Department, University of Michigan

Monday

9:00	Continental Breakfast	3201 AH
9:30	Welcome - Al Taylor	
	Overview of the Week - Pat Shure	
	Overview of the week and introduction of the staff	
10:00	Philosophy of the Course - Mort Brown, Bob Megginson	3201 AH
	Seven minute videotape outlining Michigan Calculus followed by a discussion of the philosophy of the course.	
	Profile of Students in the Introductory Math Courses - Pat Shure, Bob Megginson	
11:00	Distribution of text, calculators and materials*	
12:00	Lunch break	
1:00	Creating an Atmosphere for Learning - Bob Megginson	3201 AH
	Demonstration of a sample "first-day" of class followed by a discussion of what is important to get accomplished including creating an atmosphere for learning.	
2:00	Lecturing vs. Explaining - Donatella Delfino, Moira McDermott	3201 AH
	Discussion of teaching for understanding. Participants view and critique two video clips of explanations in math classes .	
3:00	Teaching with the TI82 Calculator - Ed Goetze*	3201 AH

Assignment --Prepare a 5-minute talk for the Tuesday morning session using the observation sheet to guide you. Consult the Group and Topic (GAT) list for your topic and room. Note the time and room number for your group.

Assignment --Start thinking about your first-day handout. A draft will be due on Thursday.

Tuesday

9:00	Group I: Computers, Library*	
	Introduction to computers and the library system at UM	
	Group II: 5-Minute Individual Practice in Giving Explanations	MLB rooms
	Instructors are videotaped teaching a lesson to their peers and get feedback.	
10:30	Group I: 5-Minute Individual Practice in Giving Explanations	MLB rooms
	Group II: Computers, Library*	
12:00	Lunch Break	
1:00	Using Questions to Improve Thinking - Dale Winter, Beverly Black	2003 NS
	Discussion of interactive lecturing, asking questions, and other methods for getting students to think (think-pair-share, writing, etc.). Viewing and discussing of video clips of actual classroom scenes where instructors are using questions to get students to think.	
3:00	Grading System - Pat Shure, Moira McDermott, Bob Megginson, William Cherry	3201 AH
	Discussion of the grading system and how to communicate the grading system to students.	

Assignment--Prepare a 7-minute overview of an assigned section. Ask at least one question where you get a response. Be prepared to answer simulated student questions on that section of the text and any of the associated homework problems. Consult the Group and Topic (GAT) list for your topic and room. Note the time and room number for your group.

(over)

*These sessions are University specific and an outline of the session is not included in this publication.

Wednesday

9:00 Individual Practice in Asking and Answering questions — MLB rooms

Instructors give a 7-minute overview of an assigned section, asking at least one question. For 10 minutes they answer questions on the material or on that section of the book.

12:00 Lunch Break

1:00 Cooperative Learning - Pat Shure, Beverly Black — 2003 NS

Discussion of why student-focused teaching is important. Discussion of video clips of actual classroom scenes where instructors are facilitating cooperative learning. Discussion of ways to set up tasks, facilitate learning and come to a closure.

3:00 Lecturing (How Long and When) - William Cherry, Ed Goetze — 2003 NS

Discussion of how lecturing fits into the course.

4:00 Faculty: Orientation by Associate Chair - Paul Federbush* — 2003 NS

Assignment--Prepare a 15-minute lesson using a short exercise to get all of the "students" in your group actively involved and thinking. Use the "Guideline for Analyzing a Cooperative Learning Exercise" to prepare your lesson. Consult the Group and Topic (GAT) list for your topic and room. Note the time and room number for your group.

Assignment --Write up a draft of your first-day handout to hand in Thursday at 9:00 a.m. Be sure to include your grading policy.

Thursday

9:00 Individual Practice in Using Cooperative Learning — MLB rooms

Instructors give a 15-minute lesson using active learning methods to a small group of peers and receive feedback.

12:00 Lunch Break

1:00 Homework Teams - William Cherry, Dale Winter — 2003 NS

Demonstration of forming homework teams. Viewing and discussion of videotapes of homework teams in action. Discussion of how to form groups, when to form groups, developing and communicating ground rules,.

Friday

9:00 Making Groups Work - Moira McDermott, Donatella Delfino — 3201 AH

Demonstration and discussion of the fishbowl technique to help students learn how to work together in groups. Other methods for getting students to work together will also be discussed.

11:00 Grades, Ethics, Values - Dale Winter 3201 AH

12:00 Lunch Break

1:00 Planning and Managing an Interactive Class - Pat Shure, Bob Megginson — 3201 AH

Tips on planning and managing an interactive class, schedules for first week, other managing issues (e.g., absences, grade complaints, late adds/drops).

2:30 Stump the Staff* — 3201 AH

Time to ask all the questions that haven't been answered during the week.

3:30 - ? Party at Al Taylor's house. Bring a swimming suit and towel if you want to swim.

Session Descriptions

Course Philosophy

Goals for the Session

- Familiarize instructors with course goals.
- Preview the key features of the course.
- Indicate how the course differs from a traditional course.
- Acquaint instructors with the composition of the population of students taking the course.

Outline of the Session

This session is simply a talk to give the instructors an overview of the course: the Department's goals for the course, the key features, and the makeup of the student population.

State the reasons for restructuring the course.

Briefly talk about calculus reform nationwide--how students were failing in the traditional course, and how the advent of technology caused a widespread reexamination of the introductory calculus courses.

Present the course goals.

Show an overhead of the course goals and briefly discuss them. (At Michigan we show the 8-minute introductory tape before we discuss the goals.) If your goals are similar to those at Michigan, point out that much more is expected from the course than having students learn calculus skills. Following are the goals of Michigan Calculus:

Course Goals

Establish constructive student attitudes toward math:
1. interest in math
2. value of math and its link to the real world
3. the likelihood of success and satisfaction
4. the effective methods to learn math

Strengthen students' general academic skills:
1. critical thinking
2. writing
3. giving clear verbal explanations
4. working collaboratively
5. assuming responsibility
6. understanding and using technology

Improve students' quantitative reasoning skills:
1. translating a word problem into a math statement and back again
2. forming reasonable descriptions and judgments based on quantitative information

Develop a wide base of calculus knowledge:
1. understanding concepts
2. basic skills
3. mathematical senses (quantitative, geometric, symbolic)
4. the thinking process (problem-solving, predicting, generalizing)

Improve student persistence rates:
1. students continue taking math and science
2. more students become math majors

List the key features of the course.

Show another overhead listing the key features of the course. Briefly describe each feature and tell the instructors that each of these features will be described more fully in upcoming sessions. Following are the key features of Michigan Calculus:

Key Features of Michigan Calculus

Syllabus
The new syllabus stresses the underlying concepts and incorporates challenging real-world problems.

Textbook
The textbook emphasizes the need to understand problems numerically, graphically, and through the traditional algebraic approach (the "rule of three") as well as by writing explanations of the work.

Classroom atmosphere
The classroom environment uses cooperative learning and promotes experimentation by students.

Team homework assignments
A portion of each student's grade is based on solutions to interesting problems submitted jointly with a team of three other students.

Graphing calculator technology
TI-82s are used throughout the introductory courses.

Student responsibility
Students are required to read the textbook, discuss the problems with other students, and write full essay answers.

Discuss the student demographics of the course.

Give a brief overview of the students who are taking the course so that instructors will understand who they will be teaching. This is specific to your department. You might include: how the course fits in with other courses offered by the department, what other courses are available to students, the number of students who will be enrolled in all of the sections of the course, the size of individual classes, and the background of the students (what courses they may have had, what their intentions might be--majors, etc.)

Equipment
Overhead transparency projector
VCR and monitor (for VHS tapes)

Material: (see Appendix I for examples)
Overheads of course goals and key features
Introductory video tape

Creating an Atmosphere for Learning On the First Day of Class

Goals for the Session

- Get instructors planning and thinking about their first day of class.
- Help instructors learn how to set an atmosphere for learning in a class that uses cooperative learning.
- Provide opportunities for instructors to get acquainted with each other.
- Provide an opportunity for instructors to see (and participate in) two different models of first day of class.
- Give instructors two examples of first-day-of-class handouts.

Outline of Workshop

Introduce the session.

Outline the goals and schedule of the session.

Demonstrate a first day of class.

Two experienced instructors (one faculty and one TA if possible) demonstrate a portion of their first day of class pretending the new instructors are their students. The demonstrations should be quite different in order to open up possibilities for the new instructors. If possible, have the first instructor act just as he/she would during the first day of class when the students (instructors) enter the room (greeting them, starting to get to know them, having them sit where he/she wants them to sit, etc.).

Between the two examples the following should be demonstrated:

1. A way for the students to get acquainted with each other
2. Learning and using names
3. Getting students actively involved in a cooperative learning task
4. Communicating goals of the course
5. Gathering information about the students
6. Using a first-day-of-class handout

After both experienced instructors get through demonstrating portions of their first day, have them give a short talk on why they do what they do on the first day. Give new instructors a chance to ask questions.

Summarize what needs to be accomplished on the first day.

Give a short talk on what needs to be accomplished and why it is important. (You might use an overhead to emphasize important points.) Refer back to the demonstrations for examples. Following is an outline of points to cover.

Establish course goals.

Start learning and using students' names.

Get students acquainted with each other. Students are more likely to work together and take risks if they know each other.

Get students acquainted with you. What kind of an instructor will you be?

Gather information about the students. The quicker you get to know students and their backgrounds the easier it will be to facilitate active learning in the class. This information will also be used to help you form homework teams.

Provide students with important information on how the class will be conducted. Much of this will be on a first-day of class handout.

1. grading policies
2. tests--ground rules
3. what they are to bring to class
4. format
    - -cooperative learning in class
    - -homework teams
    - -fewer and more challenging problems
    - -stress learning, not grades
    - -emphasize expectation that they will read the book

Get students involved in a group activity. This helps set students' expectations early that they will be required to participate and work in class.

Give a clear statement of the assignment.

Wrap-up
Encourage questions and/or concerns from new instructors.

***Assignment for instructors**: Start thinking about your first-day handout. Other sessions will address some of the things that should go into a handout. You will be asked to hand in a draft of your first-day handout to the workshop staff on Thursday.*

Equipment

Overhead transparency projector

Material (see Appendix II for examples)

a. Two first-day handouts (copies for each instructor)
b. Student Data Sheet (a copy for each instructor)
c. Overhead: "What Needs to be Accomplished"

Lecturing vs. Explaining

Goals for the Session

- Get instructors thinking about what it takes for students to learn and understand a concept rather than focusing solely on their own presentation (become student-focused rather than teacher-focused).
- Provide the basics of good lecturing methods including getting instructors to start thinking about lecturing as explaining.
- Give instructors an opportunity to observe the teaching of others and give constructive feedback, keeping in mind what it takes for students to learn and understand.
- Provide direction for the subsequent five-minute practice teaching session.

Outline of Workshop

Introduce the session.

Outline goals and schedule of the session.

Give a short presentation on how to give a good lecture.

Following is an outline of a short presentation on aspects of giving a good explanation.

Promoting understanding. Think of lecturing as explaining.

1. Put definitions of lecturing and explaining on overhead transparency projector.

   Definitions from Webster's Dictionary.

   Lecture:
   1. An informative talk given before an audience, class, etc., and usually prepared beforehand. 2. A lengthy scolding.

   Explain:
   1. To make clear, plain, or understandable. 2. To give the meaning or interpretation of; expound.

2. Think of anything you present to your class as an explanation instead of a lecture.

3. Focus on what it takes for students to learn and understand a concept, rather than just how you will be presenting the concept (student-focused rather than teacher-focused).

4. Prepare!

Connecting

1. Speak to specific students in turn, making eye contact.
2. Speak loud enough for everyone to hear (even when you are unsure).
3. Use handouts to help communication.
4. Prepare!

Writing on the Board

1. Whatever you put on the board students will copy.
2. Write clear, big, and stand aside.
3. Don't use the bottom of the board.
4. Organize the board so that someone coming in late can tell where you are and where you have been.
5. Make sure your notation is consistent with the book.
6. Say something twice and write it once.
7. Erase thoroughly.
8. Prepare!

Motivating

1. Show enthusiasm.
2. Show respect.
3. Have patience.
4. Prepare!

Introduce the "Guide for Analyzing Explanations."

Introduce the "Guide for Analyzing Explanations" and describe how it will be used.

Have instructors read the Guide.

Explain that the Guide will be used to give feedback on their five-minute practice teaching session on Tuesday.

Explain that right now they will be "visiting" a class by way of the video camera. The instructor is giving an explanation to his students. They are to use the Guide to focus their viewing of the class.

As they view the tape have instructors jot down notes on:

1. Is there anything the instructor does that would be particularly useful in helping students understand the material?

2. Is there anything that might get in the way of learning the material? Is there some place that students could become mixed up?

Introduce the video clip and view the tape.

For this activity there are two video clips on the trigger tape video under "Teaching for Understanding." Choose the one that best fits your needs.

Video clip 1: Pat Nebel
We join this class as the instructor is introducing how to find areas.

Video clip 2: Doug Shaw
We join this class immediately after a student tells the instructor she is having problems with the logarithm rules.

Conduct a discussion on the video clip.

After viewing the tape give instructors a few minutes to think about the tape and to fill out their Guides.

Have the instructors find a partner and compare notes and discuss the two questions.

Facilitate discussion in the large group addressing the two questions. Call on 4 or 5 people by name and then ask everyone at large if there are any other comments. Use the Guide to generate other comments.

Wrap-up

Discuss the assignment for tomorrow and address any questions and/or concerns about the process.

Assignment for instructors--*Prepare a 5-minute talk for the Tuesday morning session using the observation sheet to guide you. Consult the Group and Topic (GAT) list for your topic and room. Note the time and room number for your group.*

Equipment

VCR and monitor (project on screen for a large group)
Overhead transparency projector

Materials (see Appendix III for examples)

a. "Guide for Analyzing Explanations" (a copy for each instructor)
b. Overhead: Definitions of "Lecture" and "Explain"
c. Group and Topic List (a copy for each instructor)
d. Trigger tape video, "Teaching for Understanding"

Individual Practice in Giving Explanations

Goals for the Session

- Get instructors thinking about what it takes for students to learn and understand a concept, rather than thinking about their own presentation (help them become student-focused rather than teacher-focused).

- Give instructors an opportunity to teach for understanding and receive feedback from their own self-analysis, their videotaped presentation, their peers, and the facilitator.

- Familiarize instructors with the textbook.

- Give instructors a chance to practice observing others and practice giving constructive feedback on teaching.

- Provide an opportunity for instructors to see others teach.

Outline of Workshop

Form small groups of 5 or 6 instructors with a room, video camera, and facilitator for each group. Each instructor comes prepared to teach a concept from the textbook (pages are assigned the night before). The facilitator will run the session as outlined below.

Conduct an activity to get participants acquainted.

Facilitate a quick activity that gets everyone acquainted and relaxed. For example, you might have pairs interview each other and then take turns introducing each other to the whole group. Have them include something interesting, such as a unique hobby or something that others might not ordinarily find out.

Explain the videotaping process to the instructors.

The instructors will take turns presenting for five minutes, one right after the other. Keep camera running: as the next person goes up in front, the instructor who just finished goes to the camera to videotape the next presenter. The first person to videotape is the last to present. Instructors enjoy running the video camera, and if they operate the camera immediately after their own presentation they worry less about their performance.

The instructors will time themselves. They will set the timer before they start.

Tell the instructors in the "class" to take notes as if they were students trying to remember the material for a test. Have them use a separate sheet of paper for each presenter. They should put the presenters name on the notes because they will be giving the notes to the presenter later in the session.

Begin the Videotaping.

1. Decide the order in which participants will present.
2. Make sure that each person sets the timer.
3. Videotape one after the other.

Break
When everyone has been videotaped give them a short break while you rewind the tape.

Play back the videotape and facilitate feedback sessions.

Check how much time remains and keep the feedback session on schedule so each person gets equal feedback time. Sometimes you will have to break off the discussion to do this.

Tell them you are going to play back each explanation one at a time and facilitate a feedback session for each instructor before going on to the next one.

Give everyone (including the presenter and the facilitator) a "Guide for Analyzing Explanations" sheet for each presenter.

Play the first presentation and then stop the tape. Encourage participants to write constructive comments (both positive comments and suggestions for change) on the Guide for feedback to the instructor.

Facilitate a discussion on the instruction. Begin with the good things about the instruction before addressing those things that didn't go so well. Encourage other participants to give feedback before you give feedback.

1. Ask the presenter to go first.
   What went right? Did it go as planned?
   What did she/he think worked?
   What does he/she especially want feedback on?

2. Ask participants to give feedback.
   What worked--what did the instructor do that would help students understand the material?
   Was there any place students might get mixed up?
   What could be done to make the learning better?

3. Use the analysis sheet to generate other feedback.

4. Your feedback should be intermixed with the feedback from the instructors. Emphasize those things that really worked, but also discuss possible changes that would make the presentation better. You may be able to use this as an opportunity to emphasize the goals of the course.

After the discussion is over everyone (including the facilitator) gives the presenter his or her notes and analysis sheet.

Play each presentation and go through the above process with each one.

Equipment (the following are needed for each small group):
Video camera and playback system (VCR and monitor)
Chalkboard and chalk
Timer, blank videotape

Material (see Appendix III for example)
a. "Guide for Analyzing Explanations"--enough copies so that everyone (including the presenter and the facilitator) can fill out one for each person in group.

Using Questions to Improve Thinking

Goals for the Session

- Demonstrate the difference between teaching that fosters passive learning and teaching that requires active learning.
- Continue creating the shift from thinking about what it takes to teach a concept to thinking about what it takes for students to learn a concept.
- Provide instructors with an opportunity to view and analyze videotapes of other faculty members and TAs who use methods which get students actively involved and thinking.
- Provide direction for the subsequent practice teaching session.

Outline of Workshop

Introduce the session.

Outline the goals and schedule for the session.

Introduce and show the first video clip, "Are there any questions?"

A short introduction will help the instructors focus. You might say:

> *Much of this session is about asking questions so we thought we would start out by showing a videotape of several faculty and TAs asking the most often-asked question.*

This is meant to be a humorous beginning, but after you show it you could point out that few students, if any, will ever respond--and then only the brave of heart.

Give a short presentation relating session to goals of the course.

Give a short presentation on how the goals of the course relate to the session. A sample presentation follows:

> *In this session we would like to have you think about what it takes for students to learn and understand calculus. We would like you to develop goals and methods in your teaching that focus on students' learning. Think about what you can do in class that will help achieve the goals of the course.*
>
> 1. *Improve students' thinking and analytical skills.*
> 2. *Help students gain a deeper understanding of concepts.*
> 3. *Provide students with a classroom experience that requires them to be actively involved and thinking.*
>
> *One of the most important tools you will use to get students actively involved and thinking is the use of questions. In order to get a sense of how one might go about using questions in a way that gets students involved, we will view and discuss several videotape clips of actual classroom scenes where instructors are asking questions.*

View and discuss the video clips.

For this activity there are four video clips of classroom scenes with instructors using questions to get students to think. You may wish to show a subset of these videos and save some to analyze after the term begins. It is important to give a good introduction and establish a context in which to view the tapes. Each clip is described below with questions that will focus the viewing.

Show the video clips one at a time and facilitate a discussion after each. As you discuss the tapes, focus on questioning. How do you use questioning to get all students actively involved and thinking? How much time should you leave after asking a question so that students will think? Why call on students by name? These are demonstrated in these video clips.

Note: It is quite important to start out the discussions with talking about what is working, what is effective. Too often discussions can get bogged down with minor points of negative criticism. After discussing the effective aspects of the teaching it can be productive, however, to ask if there are any concerns about the instruction.

Video clip 2: Suzy Weekes
We join this first introductory calculus class about midway in the session. The teaching assistant, Suzy Weekes, is giving a lecture about areas between curves.

Before the instructors view the tape tell them to look for the following:

1. What does the instructor do that helps students understand?
2. What does the instructor do that gets students actively involved and thinking?
3. Is there anything that the instructor does that creates an atmosphere that allows students to take risks?

Video clip 3: Tony Giaquinto
This is a video of a class near the end of the second semester. Tony Giaquinto has asked if anyone finished the arc length problem that they had started on the previous day. He wrote the problem on the board and we join the class where he is asking students for their answer to the problem.

Before the instructors view the tape tell them to look for the following:

1. What does the instructor do that gets students actively involved and thinking?
2. What does the instructor do that creates an atmosphere that allows students to take risks?

Some points to bring out in the discussion:

1. Use of students' names
2. Wait time, the instructor is a master at waiting for students to think through the problem.
3. The instructor gets students to make a stance.
4. The good natured way that the instructor handles being wrong.

Video clip 4: Al Taylor
This is a video of a class midway through a second-semester class. We join the class where the instructor is challenging the class to think about estimates.

Before the instructors view the tape tell them to look for the following:

1. What does the instructor do that encourages students to think?

Some points to bring out in the discussion:

1. The instructor uses questions that require a fairly high level of thinking.
2. He is determined not to tell them the answer--waits a long time.
3. He gives them time to discuss the questions with a partner, giving everyone an opportunity to think through and discuss possible answers.
4. He uses names, this time Mr.___________.

Video clip 5: William Jockusch

This class occurs near the end of the first term and the instructor is introducing anti-derivatives. At the beginning of class the instructor invited the students to work with other students and had them find some anti-derivatives of various functions. We join the class where he is telling them to find the anti-derivative of Cos 2x.

Before the instructors view the tape tell them to look for the following:

1. What are some of the things that the instructor does that helps students understand the concept?
2. What does he do that encourages students to think?
3. What does the instructor do that helps create an atmosphere that allows students to take risks?
4. How is this class different from the first classes that we visited?

Some points to bring out in the discussion:

1. All students are being required to think and actively participate.
2. Instead of asking for volunteers, the instructor calls on students by name.
3. The instructor doesn't let the students off the hook. If they don't know something he keeps coming back to them to ensure that they find out.

After the discussion, ask what they think the instructor's objectives were for this class period. What did he want students to be able to do by the end of class?

Note: This is a good example of discovery learning. By the time the students left the class most of them understood anti-derivatives and how to check out their answers. William's goal might have been:

> To help students understand anti-derivatives by experimentation and discovery; guess-and-check.

Assignment for Instructors--*Prepare a 7-minute overview of an assigned section. Ask at least one question where you get a response. Be prepared to answer simulated student questions on that section of the text and any of the associated homework problems. Consult the Group and Topic (GAT) list for your topic and room. Note the time and room number for your group.*

Equipment

VCR and monitor

Materials

Trigger tape video, "Getting Students to Think"

Grading System

Goals for the Session

- Provide general guidelines for developing a grading system.
- Inform instructors of the grading policies of the course.
- Show instructors how to communicate the grading policies to students.

Outline of the Workshop

Introduce the session.

Outline the goals and schedule of the session.

Give a short talk on the role of grades and grading policy.

Give a short talk on the role of grades and grading policy. Following are general guidelines that could be included:

Importance of grades--Grades are extremely important to students. Not being clear about the grading policy creates more anxiety among students than any other aspect of a class.

Grading policies must be fair--The grading policies must be seen as fair and students must know and understand them. Being fair does not mean that there is lenient grading or that grades are given away for no special reason. Make sure that you have thought through your grading policies and can give reasons for grading as you do.

Grading policies must be clearly defined--Grading policies must be clearly defined at the beginning of the semester. Students must know what the system is and how it works. The students should be told up front how the instructor will grade--it should be a matter of public record.

Grading policies must be consistent--If the instructor says that grades will be determined by a certain method then that is the method that should be used. Don't change directions in the middle of the semester.

Students must have ways of knowing how they are doing in the course--Let them know how they are doing along the way. For example, at midterm you could tell them how they are doing, "*There have been 225 points possible up to this point and you have 200 points--it looks like you are in the 'B' area.*"

Outline course grading policies.

It is extremely important for you to develop clear guidelines on course policies on grading and help instructors understand them. If there are different sections of the course, students tend to talk to each other and if it is perceived that it is more difficult to get a good grade in one section than another, there will be difficulties. It is essential for everyone teaching the course to develop some degree of consensus.

<u>*Conduct a discussion on what is important to communicate to students*</u>.

Have instructors look at samples of first-day-of-class handouts to see how policies on grades and grading are communicated.

Conduct a discussion about what things are important to include in a handout.

Assignment for instructors
Develop your first-day handout describing your grading policies, your expectations, and course information. The handouts are due on Thursday.

Material: (see Appendix II for examples)
a. Samples of first-day-of-class handout (copies for each instructor)

Individual Practice in Asking and Answering Questions

Goals for the Session

- Give instructors practice in asking questions that both get students thinking and get a substantial response.
- Help instructors learn effective methods for answering student questions.
- Give instructors a chance to practice thinking on their feet.
- Provide another opportunity for instructors to get acquainted with the book.

Outline of Workshop

Form small groups of 5 or 6 instructors with a room and facilitator for each group. Each instructor comes prepared to answer questions on a section of the textbook (pages and rooms are assigned on the GAT list that they received the first day). Each facilitator will run the session as outlined below.

Introduce the session.

Make sure everyone knows each other. Outline the goals and schedule of the session.

Explain the process to the instructors.

The instructors will take turns doing the following:

> Each instructor will take no more than 7 minutes outlining the section assigned to him/her including asking at least one question and getting a response. The instructor then asks the group for questions on that section of the book or the homework problems from that section.
>
> They will time themselves. They should set the timer for 7 minutes when they start.
>
> The students (members of the group) will ask the instructor questions and the instructor will answer the questions to the best of his/her ability.
>
> The group will then give feedback to the instructor.

Facilitate the activity.

This activity requires you, as the facilitator, to keep it going and to keep good tabs on the time to ensure everyone gets a turn and that it doesn't drag. Be prepared with questions of your own to use if others don't come up with questions.

After each instructor presents for 7 minutes and answers questions for five to eight minutes, facilitate a discussion on the instructor's introduction and how he/she asked and answered questions. Begin with the good things first before addressing those that didn't go so well.

Some of the things to look for:

What kind of questions did the instructor ask? Did they work? Why or why not?

Did the instructor make sure everyone heard the question before trying to answer it?

Did the instructor ever check to see if another "student" could answer the question before he/she answered the question?

Was the answer clear and understandable?

Was the answer complete?

Did he/she check with the "student" to see if the answer was adequate?

How did the instructor handle questions he/she didn't know? (Did he/she get flustered or say "I don't know" and be willing to get back to the "student" later.)

Was the instructor respectful of "students?"

Use every opportunity to point out good methods for answering questions. Discussion should also include guidelines for when the instructor should answer the question and when he/she should encourage another student to answer the question.

Equipment (the following should be available for each small group)
Timer
Chalk and chalkboard

Using Cooperative Learning in the Classroom

Goals of the Session

- Demonstrate the value of using cooperative learning methods.
- Give instructors a clear sense of the mechanics of using cooperative learning in class.
- Give instructors an opportunity to view and analyze videotapes of other faculty members and TAs using collaborative learning methods in class.
- Help instructors learn to choose appropriate tasks for collaborative learning exercises.

Outline of Workshop

Introduce the session.

Outline the goals and schedule of the session.

Give a short presentation.

Give a short presentation connecting cooperative learning to the goals of the course. Using cooperative learning will help reach the following goals:

Improve students' thinking and analytical skills--Working in groups helps students clarify their thoughts, and they become better able to think creatively. Group work makes maximum use of each student's differing experience and knowledge. They will learn from each other.

Promote a more flexible approach in solving problems--When students get a chance to work together, they get a chance to see several ways to approach a problem. This is magnified when they see the results from other groups.

Help students to learn to work effectively with other people--Future employers are giving an increasingly strong message that they need students who are problem solvers and are able to work in groups.

Working in class with the instructor to guide them will help prepare students for working in homework teams--How they work inside of class determines how they work outside of class.

Provide a deeper understanding of key concepts--If students struggle together to figure out problems, they gain a deeper understanding of the concepts than they would if they were told the steps. When students learn to talk about mathematics they clarify their own ideas.

Reduce math anxiety and phobia--Research shows that working in groups makes math class less discouraging for anxious students. Students have gang confidence. They are more comfortable in not knowing something if they realize that others are in the same situation.

Experienced instructors talk about their use of cooperative learning.

Have one or two faculty members who have used cooperative learning in a math class give brief talks about their experience. It helps if at least one presenter really didn't believe it would work, and possibly had some difficulties in making it work at first, but came to value the process. Include discussion on what they think the value of cooperative learning is to them and to the students.

View and discuss videotapes.

For this activity there are four video clips of classroom scenes where the instructor and students are engaged in a cooperative learning task. (You may wish to show a subset of these clips for this session and show and analyze the others after the instructors have had a chance to gain some experience in conducting cooperative learning exercises.)

Have instructors form themselves into groups of 4-5 (give them time to introduce themselves to each other).

Have each group choose a reporter.

Pass out the "Guideline for Analyzing a Collaborative Learning Exercise" (Appendix IV-a) and have instructors read and use the guide to inform their viewing.

Introduce and show the first video clip.

Have the groups address the following questions in their groups

1. What did the instructor do that made it work?
2. What about the task made it a good task? Or was it a good task?
3. What did the instructor do to get students to extend their thinking?
4. Were there any concerns about the class?

Get groups to report their discussions to the whole group.

> NOTE: Instructors tend to need help in introducing a task and bringing the material together at the end. There are good examples of each in these tapes. Make sure these aspects of using cooperative learning are discussed.

Show the other clips one at a time and repeat the process. For the third and fourth clips you might facilitate the discussion in the whole group.

Generally, when the instructors discuss the video clip in small groups first, a better discussion will happen in the large group. You can point that out and relate it to what happens in the classroom.

Video clip 1: Chris Towse
We join this first-semester introductory calculus class about three months into the term. Chris Towse is in the process of setting up a cooperative learning task.

Some points to bring out in the discussion:

1. What the instructor did to set up the task.
2. Why it is important to spend time setting up the task for students.

Video clip 2: Lester Coyle
In this class Lester Coyle is giving the students an opportunity to recap what they studied the day before by giving them two problems to work on in groups. The problems require students to find the integral using the area above the curve. The activity took about 27 minutes of the class time. In the video we will watch video clips of the exercise showing how Lester handles three portions of the cooperative learning activity: setting up the activity, working with the small groups, and bringing the material together at the end.

Some points to bring out in the discussion:

1. Forming groups by counting off helps students get to know other students in the class and gives them an opportunity to be exposed to the thinking of several students.
2. The instructor was very respectful of students ideas.
3. The instructor constantly encouraged and praised student work.
4. The instructor was constantly asking questions that extended students' thinking.
5. The instructor wrapped up the activity by getting students to report their solutions and by adding to the solution himself to help students learn.

Video clip 3: William Cherry
This video clip is of a short exercise on finding functions which are their own inverse. We see the whole exercise from beginning to end. It takes about 6 1/2 minutes. In this video clip have the instructors pay particular attention to the students and what they are doing.

Some points to bring out in the discussion:

1. Often it will take students a while to settle down and work on the task.
2. Some tasks require less talking and discussion than other tasks.
3. This exercise might be a good one to do in pairs rather than groups.

Video clip 4: Victoria Pambuccian
In this class Victoria Pambuccian is having students find a function that is the derivative of an exponential function. The activity took about 35 minutes of the class time. We view four short clips of the activity. As instructors view the video have them focus on how Victoria manages the activity.

Some points to bring out in the discussion:

1. Focus on how she manages the activity.
2. Discuss the appropriateness of the problems chosen.
3. The problem is outlined on the board when the students came in.
4. The TA had two problems on the board but only had groups report on one of them (the second problem being an activity to keep fast groups busy while slow groups finished the first problem).

Conduct an activity on choosing cooperative learning tasks.

The goal for this activity is for instructors to learn how to choose good cooperative learning tasks for use in class.

Pass out 4 or 5 different cooperative learning tasks (one set for each group) that could be given to students. Have the groups discuss and evaluate them.

1. Are they appropriate as an in-class collaborative learning task?
2. Why or why not?
3. If a task is appropriate to use in class, identify when it could best be used.

When the groups finish, facilitate a discussion on each task with the whole group.

Wrap up the session.
Encourage questions and concerns about using cooperative learning.

Assignment for Instructors*--Prepare a 15-minute lesson using a short exercise to get all of the "students" in your group actively involved and thinking. Use the "Guideline for Analyzing a Cooperative Learning Exercise" to prepare your lesson. Consult the Group and Topic (GAT) list for your topic and room. Note the time and room number for your group.*

Equipment

VCR and monitor (project on screen for a large group)

Materials:

a. "Guide for Analyzing a Cooperative Learning Exercise" (one copy for each instructor--see Appendix IV for an example)
b. Trigger tape video, "Cooperative Learning"

*You many want to postpone this activity until after the term begins.

Lecturing (When and How Long)

Goals for the Session

- Help instructors gain a sense of how to balance their class between lectures and other activities.

- Give instructors helpful information on getting the students to read the book.

Outline of the Workshop

Introduce the session.

Outline the goals and schedule of the session.

View and conduct a discussion of clips of videotapes.

There are two video clips to show in this session. Before the instructors view each of the clips tell them to try to figure out how these lectures fit into each class. After viewing a clip, conduct a discussion about the value of the short lecture.

Video clip 1: Doug Shaw
In the first clip Doug Shaw recaps what students have been working with for the first half of the class and continues on to give a brief overview of what is yet to come. It is about 10 minutes long and is the total time spent lecturing in that class period.

Video clip 2: David Kauch
The second clip is a five-minute lecture with the instructor reviewing a concept just before it is to be used in a cooperative learning task.

Have a panel discussion on when and how long to lecture.

Have two or three experienced instructors give a brief presentation on when and how long they might lecture. Give the new instructors a chance to ask them questions.

Give a short presentation on getting students to read the book.

Have the instructors quickly read "Getting Your Students to Read the Book." Follow this with a short presentation on getting students to read the book. We have found there tends to be a negative correlation between carefully covering all the material in lectures and how much the students read the book. We have also found that students who take responsibility for reading the book report that they like the book. It is important for the instructors to be positive about the book and to help students learn how to learn from the book.

Wrap-up of the session

Briefly review appropriate times to lecture:

> To clear up common misunderstandings which surface during in-class group exercises.
>
> To let students know what to expect in their reading--a brief overview of the reading to get them started,

To give a quick overview of the information students need to work on a problem (example in tape),

To go over the difficult parts when examples from the book are both important and difficult,

To briefly summarize what they have learned that day at the end of the class period.

Some general guidelines:

1. Don't tell them what you can show them.

2. Don't show them what you can get them to discover themselves.

3. Don't lecture more than 10 or 15 minutes at a time.

3. Never lecture as if the students haven't read the book.

4. You might lecture more at the beginning of the semester (giving students a little of what they are used to--but still no more than 10 or 15 minutes at a time) and gradually wean them to less.

Equipment

VCR and monitor (project on screen for a large group)

Materials

a. Handout "Getting Students to Read the Book" (a copy for each instructor--see Appendix V-a)
b. Trigger tape video, "Lecturing: When and How Long"

Individual Practice in Using Cooperative Learning

Goals for the Session

- Help instructors understand the value of using methods in their teaching that require active learning on the part of the students.

- Give instructors a chance to practice using methods for getting students actively involved and thinking and provide them feedback on their teaching.

- Provide another opportunity for instructors to get acquainted with the book.

- Broaden the instructors' view of activities that can be used in a math class to get students actively learning the material.

Outline of Workshop

Form small groups of 5 or 6 instructors with a room and facilitator for each group. Each instructor should come prepared to give a 10-15 minute lesson using some portion of that time to get all students in the class actively involved and thinking through a cooperative learning task. Each facilitator will run the session as outlined below.

Introduce the session.

Make sure everyone knows each other. Outline the goals and schedule of the session.

Explain the process to the instructors.

The instructors will take turns doing the following:

> Each instructor will give a lesson from their assigned chapter, both teaching for understanding and providing a chance for the "students" in the class to become actively involved and thinking in a cooperative learning task.
>
> Instructors will time themselves. They should set the timer for 15 minutes when they start.
>
> The group will give feedback to the instructor immediately after her/his lesson.

Facilitate the activity.

This activity works best if the facilitator moves the session along and keeps good tabs on the time to ensure everyone gets a turn and that it doesn't drag. There should be enough time for 5-10 minutes of feedback for each instructor. Begin with the effective aspects of the lesson before discussing changes that could make the teaching better.

Follow the general outline that you did with the session on "Giving Explanations":

1. Ask the presenter to go first.
2. Encourage feedback from participants
   What worked?
   Was there any place students might get mixed up?
3. Intersperse your feedback with feedback from the other instructors.

Use the "Guide for Analyzing a Cooperative Learning Exercise" to direct the feedback to the instructors:

1. How did the instructor ensure that all students were doing the thinking?
2. Did the instructor give clear, detailed instructions for "students" to do the assigned task?
3. Did the instructor create an opportunity for individuals to share their results with others in the group?
4. Was the task appropriate, challenging, interesting, with a clear and useful purpose?
5. Would the activity enhance students' understanding of the material?
6. Did the instructor bring the material together at the end?
7. Give feedback on the following (see Guide for Analyzing Explanations):
   States topic clearly at the beginning.
   Lesson is organized.
   Emphasizes the main idea.
   Relates material to something students already knew.
   Gives suitable examples.
   Material is appropriate to level of beginning undergraduates.
   Instructor shows enthusiasm.
   Board work is clearly organized and readable.
   The pace is appropriate.

Use every opportunity to help instructors think of different ways that students could be involved in class. Where appropriate, share some of your experiences and what has worked for you.

Wrap-Up
Give instructors a chance to ask questions or express any concerns.

Equipment (for each small group)
Timer
Chalkboard
Chalk

Using Homework Teams

Goals for the Session

- Help instructors understand the value of using homework teams.
- Give an example of how to organize homework teams.
- Help instructors recognize when teams are working well together and diagnose malfunctioning teams.

Outline of Workshop

Demonstrate a possible class introduction to homework teams.

Before this session, divide the instructors into "homework teams" with four instructors in each team. Make the teams as diverse as possible (gender, race, ethnicity, knowledge, etc.) and assign each team a number. Place numbers around the workshop room to indicate where each team should meet.

When instructors come into the room, give them a list of teams and instructions on how to proceed (see example in Appendix VI-a). You may have to encourage some instructors to sit with their teams and participate in the activity.

After they have had time to form their groups and get acquainted with each other continue to demonstrate a possible introduction to homework teams.

> *Pretend for a minute that you are students and that I am the instructor. You have now become acquainted with three other people who you will be working with over the next month as a homework team--after a month we will change teams. The work that you do together in your homework teams will help you reach several important goals of this course.*
>
> 1. *Engaging with others to solve difficult real world problems will help you understand calculus at a deeper level, gain a more flexible approach in solving problems, and improve your thinking and analytical skills.*
>
> 2. *Working with your team will give you a chance to learn how to present your ideas in a clear, thoughtful, organized way. This will help you solidify your ideas--make it so you do understand. This isn't an easily learned skill, but it is an important aspect of college-level thinking in all courses.*
>
> *There is experimental evidence that if a group is working well both weaker students and stronger students will benefit.*
>
> *In your group you will each have a role. Look at the handout listing the groups. On the bottom of the sheet is a list of the roles that each of you will take over the next four weeks. For example, Jenny is #1 in her group so the first week she would be the "scribe" and the second week she would be the "clarifier" and so on. Please find the role you will play this week. (Give them time to find their roles.)*

Now take out the handout that describes the different roles. Please read this tonight and become very clear about what the person in each role will be doing. Also read the page on cooperative behaviors that will help you know how to work in your team and look at the evaluation form that team members will fill out at the end of your time together. (See Appendix VI-b, VI-c, and VI-d for examples.)

Tomorrow we will be spending some time discussing in more detail how you can make the homework teams work for you and you will get your first team homework assignment. Please read these sections tonight and come in with questions tomorrow. For now I would like to give you a chance to work together in class with your team. ...

Facilitate a discussion of the demonstration.

Facilitate a discussion of the demonstration. This works best if done by a staff member other than the one who demonstrated the process.

Discussion might include:

1. Outline what was accomplished in the demonstration (see example overhead in Appendix VI-f)
   - a way to quickly get students into their homework teams so that it doesn't take too much class time.
   - an activity where students can get acquainted with each other.
   - a chance for teams to work together in class.
   - a way to help students begin to see how working in teams can help them as students.
   - a way to help students begin to learn their responsibilities to the homework team and also how they can make the group work.

2. Emphasize that the demonstration was one way to set up homework teams (with indication there are other ways).

3. Discuss when to set up teams. At the University of Michigan we encourage instructors to set up teams on the second day of class.

Give instructions on assigning students to teams.

Describe a method they might use in forming their teams, for example:

Use the student data sheets gathered on the first day to form teams (see first day of class data sheet in Appendix I).

Put together students who live in approximately the same location on campus to make it easier for them to get together.

Make the groups as diverse as possible (gender, race, knowledge, etc.) given the information you have. There is evidence that everyone gains when there is a diverse group. There is also evidence that if students from diverse backgrounds have a good experience working together it creates more tolerance and respect for those who are different than themselves. (For the first teams don't worry about this too much. You will be better informed about each student the second time you make team assignments.) Until you know your students well, however, do not put one woman with two or three men.

Panel of experienced instructors discuss homework teams.

Two experienced instructors (one faculty and one TA) explain how they set up the teams, how they made the teams work, and what concerns and problems they had. Leave enough time for new instructors to ask questions.

Give a mini-lecture on what students say about homework teams.

What students say about homework teams:--During the past three years this course has gathered midterm feedback from students. Generally, students say that working in homework teams helps them learn the material. Following are representative student comments about homework teams:

-We help and learn from each other--it helps us understand better.
-We get the whole picture from assembling the bits that each individual understands.
-Homework teams keep us motivated (otherwise some of us would just quit).
-When someone doesn't understand there are people right there to help.
-Team homework helps us explore options.
-It helps to talk to others about homework problems in order to cement ideas and clear up any confusion.
-Homework teams give us a chance to ask questions.
-It helps us understand the concepts more when we explain them to someone else.
-It's easier to talk to each other than to talk to the professor.

There are also concerns and problems that students have with homework teams. Representative comments include:

-Team homework requires too much time.
-We need more feedback on team homework--please go over in depth those problems that were wrong in many teams.
-Some people do all the work, some don't show up.
-Some homework teams work better than others and individuals feel disadvantaged.
-Dependency on others is sometimes detrimental when the others don't care about their grades.
-The scribe learns the most--others play a more passive role.
-The groups don't always work because of personality conflicts.

As you can see, while many students really like and value homework teams, some students have misgivings and these students tend to be vocal. The three things they like least tend to be the time it takes, freeloaders, and team grades. We have found that the more the students give team work a chance the more likely they will find it valuable.

View a video clip and facilitate a discussion of two homework teams working.

As the instructors view the clips have them focus on how the teams are working together. Within their groups of four, have them compare the two teams and decide whether or not each is a functional team or not, and why. Facilitate a discussion in the larger group.

Wrap-up

Give the instructors a chance to ask questions and voice concerns.

Equipment

VCR and monitor (project on screen for a large group)
Overhead projector

Materials (see Appendix VI for examples)

A copy for each instructor of the following:

a. List of teams and roles for first four weeks
b. Homework Teams: Rules and Roles
c. Cooperative Behavior
d. Team Evaluation Form
e. Checklist for Homework

"Student Data Sheet" for first day of class (see appendix II-b)

Numbers to put on the tables
Trigger tape video, "Homework Teams"
Overhead: "What is Important in Forming Teams" (see appendix VI-f)

Making Groups Work

Goals for the Session

- Make instructors aware of the dynamics of group work.
- Demonstrate an exercise that instructors can use to help students learn how to work in groups.
- Give instructors an opportunity to feel what it is like to participate as a student in a collaborative learning task and to get feedback on their participation.
- Give instructors an opportunity to observe and give feedback to others working in a group.
- Provide methods for making homework teams function well.

Outline of Workshop

Introduce the session.

Outline the goals and schedule of the session.

Conduct a group dynamics exercise.

This exercise is designed to help instructors become aware of group dynamics during a cooperative learning exercise.

Have instructors form groups with 6-8 in each group.

Have each group split into two equal teams with an "A" team and a "B" team.

Describe the activity to the group.

1. A-Team will collaboratively work together to find a solution to a problem.
2. B-Team will us the "Group Observation Worksheet" to guide their observations of the A-Team group dynamics and record what they see. They do not participate in the problem solving activity nor try to work the problem on their own. They will be giving feedback to the A team from their worksheets.

Facilitate the activity.-

1. Set up and introduce the problem very carefully (as you would want the instructors to set up a problem for the students in their class).
2. Give a cooperative task to the A-Teams (usually a difficult real-world problem written on sheet of paper--one sheet per group).
3. A-team will collaboratively work together to find a solution to the problem.
4. The B-team will use the "Group Observation Worksheet" to guide their observations of the A-team group dynamics and record what they see. They do

not participate in the problem solving activity nor do they try to work the problem on their own.

5. When most of the A-teams have reached a solution call time.

6. In their small groups, have the B-teams share their observations of the group dynamics with the A-teams.

7. Repeat the process with B-teams being the problem solvers and the A-teams being the observers.

Facilitate a discussion with the whole group. Use the "Group Observation Worksheet" to guide the discussion.

1. Ask instructors to describe to everyone what was working.
    Did someone do something to ensure everyone understood?
    How did the group keep focused?
    Did the group use a mechanism to understand the problem?
    How did the group reach consensus?

2. Ask instructors to describe something that was not working.
    Did anyone feel excluded from the process?
    Did anyone dominate?

In response to problems within the groups get instructors to come up with some things they might do in class to address similar problems and help groups work better.

Have a brief discussion of gender dynamics.

Usually during the above activity, some of the men in small groups will dominate the activity and some of the women will feel left out and not listened to. This is a good time to discuss the data on gender differences in participation and ways instructors might counteract this bias.

Discuss how instructors might use the above exercise in class.

Discuss with the instructors how they might use this exercise in class to help students develop a sense of how to work effectively in groups. Getting a chance to see other students solving a problem with a checklist to focus their observation gives students insights on how to make their groups work.

Describe to the instructors how you set up the exercise including:
1. the criteria for the selection of the tasks
2. how the problem was set up and introduced and why
3. why tasks were given to groups written on a paper and only one per group.

Give instructors an opportunity to ask questions and share concerns about the process.

<u>*Facilitate a discussion on other ways to help teams learn to work together.*</u>

Discuss in the large group different ways of helping teams to work. This could include the following:

Help students understand the ground rules for working in groups. This goes a long way in helping students learn to work together successfully. Get instructors to look at the handouts that will help students understand their roles ("Homework Teams - Rules and Roles," and "Cooperative Behavior," and the "Team Evaluation Form"--see appendix VI) and discuss how they might use them to help students become aware of the ground rules for working together in class. As you talk about homework teams stress the individual's responsibility to the group.

Have students use the "Checklist for Homework" as they prepare their homework (Appendix VI-e).

Give students a chance to start their homework in class where you can observe and give feedback to the whole class on what you see is working or not working.

Help students learn to work together effectively in class. This will carry over outside of class.

Give students time at the end of a cooperative learning task to discuss group process.

1. How did we work together today?
2. What can we do better next time?

<u>Wrap-up</u>

Give the instructors a chance to ask questions and voice concerns.

Materials (see Appendix VII for examples)

a. Group Observation Worksheets (a copy for each instructor)
b. Two problems for groups to work (one of each problem per group)

Grades, Ethics, Values: Mutual Respect

Goals for the Session:

- Provide a forum where instructors focus on ethics, values and respect for students
- Provide guidelines on cheating and how to prevent it.

Outline for Session:

Introduce the session.

Up until now in this program we have addressed many activities that help instructors reflect on their values of teaching and help them develop respect for students. This session is designed to focus further on these issues and give instructors guidelines for creating mutual respect in their classrooms.

Present a set of guidelines that experienced instructors follow.

1. Learn your students' names and use their names as early in the term as possible. Encourage the students to get to know each other. Make sure they know your name.
2. Always treat students with courtesy and listen to them carefully. Expect them to treat both you and their classmates respectfully.
3. Don't be afraid to let your students know something about you personally outside of your role as teacher.
4. Don't waste the students' time. Begin and end class promptly. Prepare thoroughly for each class.
5. When you prepare for each class, pay close attention to the reading and homework problems that you have assigned. Students get angry and frustrated when their instructor stumbles through a homework solution, or doesn't know what they are referring to in the textbook.
6. Hand back the students' tests and homework assignments promptly.
7. Meet your obligations. Keep all appointments you make with students. When you hold office hours, don't leave early assuming no one is coming. Discussions during office hours create good relations with students who then get a chance to know you more personally.
8. Don't make snap judgments about students' capabilities. The students will often prove you wrong.
9. Don't discuss a student's performance or behavior with other students.
10. Be very careful not to categorize students on the basis of race or gender.

11. As a teacher in charge of a class you must be very careful to avoid any non-professional relationship with your students. Close personal or romantic attachments are inappropriate, even when completely mutual (notice that this reads 'mutual' and not 'equal').

<u>Facilitate a discussion on cheating</u>.

Give instructors an outline of the department's policies on cheating. Help instructors know how students cheat and what to do if it happens in their class. Emphasize that it is important to prevent cheating and give them guidelines on how to do that.

Planning and Managing an Interactive Classroom

Goals for the Session:

- Describe good classroom management techniques.
- Introduce the sample daily lesson plans for the first few weeks.
- Discuss the administrative details for starting the term.
- Guarantee that instructors are prepared to begin the term.

Outline of the Session:

Introduce the session.

Previous sessions have focused on communication (what to say and how to say it) and creating a good atmosphere for learning. This session concentrates on the mechanics of organizing the information flow and planning the time.

Give a short presentation.

Give a short talk on classroom management techniques. (An overhead might be useful here.) Experienced instructors have found these techniques useful:

Organize the information flow.

1. State clearly all policies and expectations about class procedures such as tests, homework, and student behavior.
2. Announce homework far enough in advance, collect it regularly, grade it efficiently, and return it promptly.
3. Establish fair grading policies and convey them unambiguously to the students.

Plan carefully and use the time well.

1. Stay with the syllabus.
2. Schedule well in advance for quizzes, review, student presentations, etc.
3. Organize the class period. Include an overview at the beginning of class and close with a smooth summary.
4. Balance lecturing, cooperative group work and other activities.
5. Figure out ahead of time how to readjust plans when you are suddenly faced with too much time or too little time.

Introduce the sample lesson plans.

Help instructors understand that one of the most difficult things to learn about running an active class is how to apportion the class time so that lecturing and group work are properly balanced. The sample lesson plans show when and how an instructor might use a variety of active learning exercises. These group techniques can then be reused throughout the term. Make sure that they understand that the lesson plans are just suggestions to give instructors an idea of how to plan a class period.

Hand out the sample lesson plans.

Hand out the lesson plans (see Appendix X for an example) and have them read through the first few days. Solicit questions. Repeat that no one has to follow these but that the time allocation may serve as useful guidelines. (At the University of Michigan, although instructors are not required to follow the lesson plans, most of them do, and seem to appreciate that they are available. We only give them lesson plans through the first four weeks of class, although many of them would like them for the whole semester.)

Detail any term start-up procedures.

Discuss any administrative details such as: when classes begin (at U.M. classes begin at ten after the hour), how many students they can expect, where to get class lists, how students add (or drop) classes, the schedule for instructors' meetings, how to get copying done, etc.

Wrap-up

Briefly sum up the week. Encourage questions and/or concerns from the instructors. Thank them for working hard to learn some new ways of teaching. They have learned a lot. Assure them that they are ready to get off to a good start.

Equipment

Overhead transparency projector

Material (see Appendix VIII for examples)

a. Overhead: "Organize the Information Flow"
b. Overhead: "Plan Carefully and Use the Time Well"
c. Interactive Lesson Plans (one copy for each instructor)

Follow-up Workshops

Correcting and Grading Team Homework

Goals of the Session:

- Establish the difference between correcting homework and grading homework.
- Get instructors to start thinking of correcting and grading homework as giving feedback rather than "judging."
- Provide a method for grading homework.
- Give instructors a chance to become familiar with the different types of writing that students hand in.

Outline of Workshop

This workshop takes about 120 minutes. If the participants do the readings before the workshop, it will take less time.

Introduce the session.

Team homework helps students learn the material better and gives them a chance to practice writing and explaining their solutions to each other. This is the way they get feedback before it really counts. Students will learn far more and understand it at a greater depth if they take their homework groups seriously.

Because group homework is corrected and graded it gives students the feedback they need in order to learn how to explain their ideas in writing. It will also assist instructors in getting to know the students in their class better and to see what kind of work the students are capable of, where they are having trouble, and where the material might be supplemented to help their understanding.

Discuss the grading criteria for the first homework sets.

At the beginning of the term students generally don't know how to write up homework, and it is important to allow the students a grace period during which their mistakes have fewer consequences. Some instructors have found that not having the early homework count as much as it does later in the term causes less anxiety for students while they are learning to write. Remember to explain why homework counts less at the beginning of the semester. Possible grading schemes might include:

1. Early homework sets count less than do those due later in the term.
2. The first few sets of homework are graded but the grades don't count toward the total grade.

Discuss the difference between grading homework and correcting homework.

It is useful for instructors to see the difference between correcting homework and grading homework. Correcting is synonymous with giving feedback. The instructor corrects each problem as completely as possible. After correcting all of the problems, the instructor assigns an overall grade to each problem (e.g. assign a grade number between 1 and 5).

Conduct a grading exercise.

This exercise is designed to give instructors practice in using a grading scheme. It will also give them a chance to become familiar with the different types of student writing. Present a grading scheme that you are familiar with and that works for you--or describe "Doug's."

Facilitate the activity

1. Have everyone read "How Doug Shaw Marks Papers" and tell them that they will be using that grading scheme to grade four students' papers. (Or present a grading scheme that you are familiar with and that works for you.)

2. Give everyone a packet of four different student solutions to a problem. Include in the packet instructions. For example:

   This packet contains four different student solutions to problem #48 on page 91 of the Calculus textbook. Please correct these papers as if they were your students', going through and giving written corrections as you normally would, then go back and assign a grade from 0-5 (five being the best). The actual problem statement is reproduced below.

3. Instructors individually mark the papers according to the instructions.

4. Have instructors form groups of 4 or 5.

5. Have the instructors in each small group compare their grading and discuss why they graded the way they did.

6. In the larger group ask for comments from the groups. Some questions that could be asked include:
   What differences did you see?
   What issues came up?
   What was useful feedback for the students?
   What did you learn?

Discuss general rules for grading.

Have instructors read "A Few Suggestions on Grading." Discuss general rules for correcting and grading homework, emphasizing the following:

1. Keep your comments to factual statements of omissions or of what is wrong (use words like unclear, expand, explain why, how did you get from step one to step two, etc.).

2. Keep value judgments out of your comments.

3. Never use an exclamation point unless it is after a positive statement.

4. Avoid words like nonsense!!, Bull!, obvious, it doesn't make sense, etc.

5. Never underline what you write unless it is a positive statement.

Wrap-up
Facilitate a discussion about whatever issues came up. End with a discussion on what instructors can do to help students improve their writing, for example, they can:

1. Give students a checklist on what makes a good paper.
2. Hand out examples of a good team papers (if a team hands in a particularly nice paper, ask them for permission to hand it out to the class). Discuss with the class what makes it a particularly good paper.
3. Have students grade each other's papers (give them criteria to use).

Material (See Appendix IX for examples)

a. Reading: "A Few Suggestions on Grading" by Bob Megginson
b. Reading: "How Doug Shaw Marks Papers" by Doug Shaw
c. Four students' solutions to same problem taken from your own course (copies for each instructor)

Helping Students Learn to Write Explanations

Goals of the Session:

- Help instructors' understand the importance of writing.
- Give instructors a background for understanding students inability to write explanations of their work.
- Describe the steps students will go through in learning to write explanations.
- Demonstrate an activity that instructors can use with their students to assist them in learning to write.

Outline of the Session:

This session takes about 120 minutes.

Introduce the session.

Outline the goals of the session.

Give a short presentation on the value of having students write explanations

Instructors need to understand the value of having students write explanations in order for them to help students gain a sense of ease with the process. In appendix XI is a handout describing what students and instructors get out of requiring students to write explanations. Give this handout to the instructors and encourage them to read it. Then have one or two instructors, who have had experience in requiring students to write explanations, give personal experiences that have convinced them that writing is valuable.

In Appendix X there are two sheets with a student's actual solution to a problem. The first sheet (X-b) shows the student's solution without the explanation. Put this sheet on a transparency and show instructors that if one were just looking at the algebraic solution to the problem one would make the assumption that this student knew what was going on and would get nearly full credit. (There is a minor arithmetic error.) Then show the transparency of the whole solution, including the written explanation (X-c). It will make it evident that the student doesn't understand what is going on. This demonstration will convince most instructors of the value of requiring explanations.

Give a short presentation on why students complain about writing.

The requirement of using complete sentences to explain their work can create a lot of frustration and unhappiness among students. They don't understand why they are being required to do it. We often hear the remark "Why do we have to write? This is not an English class--it's a math class." "Math is numbers--not words."

Many students do not like to write explanations because they tend to not know how to do it. Most of them have never been required to write about mathematics. Some students have already had some calculus but many of them have not had experience or skill in writing explanations. It is important to help students see the value in writing explanations of their work. It is essential that instructors communicate why writing explanations is important early in the term and to reinforce the importance of explanations throughout the term. It has worked for some instructors to remind students once or twice a week (in one or two sentences), tell the students why writing is important.

Demonstrate how to approach writing with students.

We have found that it helps instructors to actually see how to introduce writing to their students. Demonstrate what they can say.

> *In this course, you will be required to explain your answers in writing. If you can explain in writing how you arrive at an answer, describing the methods you used and why, it will give you a much deeper understanding of calculus. Putting your thoughts in complete sentences will help you learn how to communicate and talk about math--it will clarify your thinking. It is a well-known fact that when someone teaches a subject for the first time, and it becomes necessary to explain ideas to someone else, the teacher comes to a much deeper understanding of the material. Explaining your answers in writing and in group work will help you in the same way.*

Describe the stages of student writing and how instructors can help students learn to write.

Put the following on an overhead transparency and discuss the stages students go through as they learn to explain their solutions in writing and what the instructors might do to help them move through these stages:

Stages of Student Writing

1. **Nothing to write**
   They cannot even conceive of what might be useful to write down.

   Ask them to explain what they are doing. Try to get them to focus on the process instead of the final solution.

2. **Description of mechanics**
   This often includes a description of every single step including the arithmetic. Often they use a two column explanation.

   Ask them to describe their thinking. Ask them why they are doing a particular step.

3. **Can't write math**
   This includes a difficulty with merging symbols and prose.

   Suggest that they try to read their answer out loud. Emphasize that symbols are just shorthand for words.

4. **On the right track**
   They're starting to get the idea.

5. **Overkill**
   Your students may never reach this stage. This includes excessive use of math jargon and an exaggerated formalism.

   Try to get students to identify the key points. Ask them to identify their audience.

At all stages, it is helpful to give students examples of good writing (these are best if it is writing done by another student).

Some helpful phrases to use when correcting student writing:

Why?
How did you know to do this?
What's the main idea?
Why this formula?
Define the symbols you have used.
This confuses me.

Conduct an exercise.

Give instructors four students' solutions to the same problem, with different levels of writing. Have them individually read the papers and decide whether or not each student has adequately explained the solution to the problem and how their writing could be improved. In small groups, have the instructors discuss each paper and the merits of the writing. Discuss each paper in the larger group. Discuss how instructors might use this exercise to help students learn to write better.

Equipment:

Overhead transparency projector

Material: (see Appendix X for examples)

a. Reading: "Why Get Students to Write?" (a copy for each instructor)
b. Overhead: Student's work showing only the algebraic solution
c. Overhead: Student's complete work showing written explanation
d. Four students' solutions to same problem taken from your own course (copies for each instructor)

Sticky Student Situations

Goals for the Session

- To provide ways for instructors to handle a number of classroom issues.
- To encourage instructors to see their colleagues as a resource when there are problems in the classroom.
- To give new instructors a sense that everyone has classroom problems.

Outline of Workshop

Introduce the session.

Outline the goals and schedule for the session.

Facilitate a problem solving session.

Have instructors form groups of four.

Each instructor chooses one of the case studies to discuss in his/her group.

The groups discuss and find responses or solutions for each of the case studies chosen by group members. (Each group will discuss four cases.)

Give the groups five minutes for each case study.

Briefly discuss each case study in the larger group, getting input from the groups that discussed the particular case.

Another possibility is to have a panel of experienced instructors address each case study.

Sticky Student Situations

1. Nicole comes to your office angry and distraught. In high school, calculus was her "favorite subject" but she feels she doesn't know what is going on in this class.

2. Jean obviously knows the material and does well on the quizzes and exams. She is very quiet in class, however, and doesn't participate. In her small group she also hesitates to put forth her ideas.

3. John doesn't show up for class very often, but when he does he tends to ask you to explain material in detail. Often the things he doesn't understand were discussed in a previous class that he didn't attend.

4. Tim doesn't like the class and is quick to complain about everything. You can feel that his negativity is spreading to other students.

5. Jeremy tries very hard but has a lot of difficulties understanding calculus. He comes in to office hours often and tries to do all the homework but he just doesn't seem to be able to do the work.

6. Nancy is very quick to answer and ask questions in class. Sometimes she leads the conversation off to a higher level than others can understand.

7. Ted and David are good friends. They are both very bright and do well on the quizzes and exams, but in class they often discuss outside events and disrupt the class.

8. During the semester Leslie hasn't done very good work, nor has she tried very much. You have tried to encourage her throughout the semester because you know she can do the work if she tried. She is going into the final with a low C. She comes into your office stressed out because she really needs a B in the course in order to transfer next semester. She wants to know what she can to make her grade higher.

9. Don comes to your office disgusted with doing group work. He feels that it is a waste of his time and he thinks he learns better on his own.

10. You have a lot of material to cover and Chris asks a key question on the preliminary material. If you answer the question adequately, however, it will make it so you can't get through what you had planned.

11. Kathy is an enthusiastic student and often answers questions without raising her hand. You don't want to dampen her spirit but she often monopolizes the conversation.

12. For the first part of the semester Keith is a model student: he is always there, he participates, it is obvious that he has done the homework, and he got high scores on quizzes and the first midterm. He suddenly starts missing class, and says nothing when he does come.

Materials

List of situations (a copy for each instructor)

Making Groups Work

Goals for the Session

- Provide an opportunity for instructors to discuss and find solutions for problems they may be having (or will have) with using groups.

Outline of Workshop

It works best to conduct this workshop after instructors have had a chance to try using groups in and outside of class. It is a rare instructor who will not be experiencing problems, but often they may think they are having more problems than others. This session will assure them that other instructors are having similar problems and that it is useful for them to discuss the problems with their peers. This session takes 60 minutes.

Introduce the session.

Outline the goals and schedule of the session.

Conduct discussions of what will go wrong and what to do about it.

Split instructors into groups of four. Have each instructor choose one of the following questions to discuss in his or her group, or they can choose another problem that they may be experiencing in their class. Participants take turns presenting their problem to the small group and the group spends 5 minutes coming up with possible responses. Use a timer and tell the groups when it is time to go on to the next problem. After everyone has had a chance to get a problem discussed in his or her small group, discuss each problem in the larger group, posting a list of responses for each question. Have experienced faculty/instructors react to the responses or add to the list.

1. What do you do when one or two students in the group don't participate?

2. What do you do when one or two students dominate the group?

3. What do you do when one group gets done way ahead of other groups?

4. When a student comes in after group work has started, how do you get him/her involved?

5. What do you do if one group is clueless or on the wrong track?

6. What do you do if all or most of the groups are clueless?

7. What do you do if the students don't work on what you want them to?

8. What do you do when the students work by themselves and don't engage with each other? What about an individual who turns away from the group and works alone?

9. Other____

Note: Instructors engaged in this activity tend to come up with more "sticks" than "carrots" as responses to the problem. Finding ways to give "carrots" can create a different atmosphere in class. Make sure enough "carrots" are covered in this activity.

Panel discussion on helping groups to work together.

Have two experienced instructors describe other methods for helping groups work successfully together. This might include:

Setting ground rules.

Encouraging groups to take the time to discuss group process.

Materials

List of problems (a copy for each instructor)

Equipment

Timer

Appendices

Course Goals

Establish constructive student attitudes toward math:

1. interest in math
2. value of math, and its link to the real world
3. the likelihood of success and satisfaction
4. the effective methods to learn math

Strengthen students' general academic skills:

1. critical thinking
2. writing
3. giving clear verbal explanations
4. working collaboratively
5. assuming responsibility
6. understanding and using technology

Improve students' quantitative reasoning skills:

1. translating a word problem into a math statement and back again
2. forming reasonable descriptions and judgments based on quantitative information

Develop a wide base of calculus knowledge:

1. understanding concepts
2. basic skills
3. mathematical senses (quantitative, geometric, symbolic)
4. the thinking process (problem-solving, predicting, generalizing)

Improve student persistence rates:

1. students continue taking math and science
2. more students become math majors

KEY FEATURES OF MICHIGAN CALCULUS

Syllabus

The new syllabus stresses the underlying concepts and incorporates challenging real-world problems.

Textbook

The textbook emphasizes the need to understand problems numerically, graphically, and through English descriptions as well as by the traditional algebraic approach

Classroom Atmosphere

The classroom environment uses cooperative learning and promotes experimentation by students.

Team Homework Assignments

A portion of each student's grade is based on solutions to interesting problems submitted jointly with a team of three other students.

Graphing Calculator Technology

TI-82s are used throughout the introductory courses.

Student Responsibility

Students are required to read the textbook, discuss the problems with other students, and write full essay answers.

Appendix II-a: Example of First-Day of Class Handout

MATH 115 --- SECTION 102 --- WINTER 1995

PAT SHURE: Office________________ Phone__________________ E-mail_________________

OFFICE HOURS: Mon. 5:00-6:00pm, Tues. 12:00-1-00pm, or by appointment.
You will be expected to see me regularly.

TEXT: ____

CALCULATOR: ____

TIMETABLE:

Exam I	Thrus. Feb.2	7:00-8:30 PM
Exam II	Thurs. March 23	7:00-8:30 PM
Final Exam	Fri. April 21	10:30AM-12:30PM

There will be no makeup quizzes or exams. I will not use class time to go over tests and grading, so please make an appointment with me to go over your tests individually.

GRADING:

Exam I	15%	Classwork	15%
Exam II	20%	Team Homework	25%
Final Exam	25%		

Team homework will be judged on:
(1) Understanding of the main ideas
(2) Clear explanations
(3) Use of the "rule of three" (if possible)
(4) Correct math; accurate calculations
Points may be added for an exceptional presentation or deducted for a poor job.

Classwork will be judged on:
(1) Daily preparation (reading, homework done)
(2) Classroom attention and participation
(3) Quiz grades
(4) Quality of your notebook portfolio
(5) Conscientious studying for exams

CLASSES: You are expected to attend every class bringing your textbook and calculator. You should read the material, think about it, and attempt the problems in preparation for class. In class we will discuss your questions and work on some of the challenging problems together. I will not always "cover" the material step-by-step in the way you may have seen before, so you will have to think hard and work with your classmates. If you need more help you can use the free tutors in the **MATH LAB.**

TEAM HOMEWORK: One complete set of the joint solutions with a cover sheet should be turned in. It should be neat (no fringes, stapled), legible, and written in correct English. Homework is due at the **beginning** of class, and I will not accept late homework.

The **cover sheet** should contain both the minutes of your meeting and the role assignments.

Scribes should **write up the solutions as if other students were the audience**. Pretend you are explaining the problem to another student who knows as much math as you do but is not in the class.

INDIVIDUAL HOMEWORK: You will have some individual problems in addition to your team problems. These will not be collected or graded. However, these problems may come up on later quizzes or tests.

Appendix II-a: Example of First-Day-of-Class Handout

Math 116
Calculus II

Instructor:	Donatella Delfino
Office:	
Mailbox:	
Office phone:	
e-mail:	
Office hours:	Mon 10:10-11 AM
	Tue 10:10-11 AM
	Thu 1-2 PM
	or by appointment
Text	
Calculator:	

1. **Grading.** Your course grade will be computed as follows:

   ▷ Quizzes: 10% (weekly, lowest score dropped)

   ▷ Class participation: 5%

   ▷ Homework-team: 25%

   ▷ First uniform exam: 15%

   ▷ Second uniform exam: 20%

   ▷ Uniform final exam: 25%

   There will be a "GATEWAY" test that you will be required to pass before the end of the semester. More will be said about it in class.

2. **Homework assignments** Your homework assignments are given in two parts. There is a daily reading assignment and a problem assignment, and a group homework assignment.

   * The group assignment is to be handed in one paper per group. It should be stapled. The first page should contain the name of the group, the full names of the group members, their roles and the report of how the homework session went. A homework set written in different handwritings will not be accepted. Every group member will receive the same grade on the group homework assignment. Homework groups will be changed approximately every month. Roles in the groups should be changed every week.

   * The daily assignments will not be collected individually. We will, however, have a weekly quiz on these problems (usually on Fridays). Please maintain a notebook of your homework solutions (for the group assignments, you may either write up your

own solutions, or include a copy of the group solution). These notebooks should be brought to the office hours, and will be collected before the end of the term. They will be returned to you. I will use my evaluation of your notebook in determining your final grade. It will be difficult for a student with a good notebook to receive a poor grade in this course. I am likely to interpret a poor notebook as an indication that a student did not make a serious effort in the course; consequently, his/her final grade will not be improved by this work.

3. **Excused absences** All missed homeworks, quizzes and exams will be given a grade "0" unless PRIOR arrangements are made. If you have advance warning of a situation that will cause you to miss a test, please discuss it with me so that you know the absence will be excused. Travel plans will never be grounds for an excused absence.

4. **Make-up quizzes** Since the lowest quiz score will be dropped, no make-up quizzes will be given.

5. **Attendance** Regular attendance is important. I will notice if you don't attend regularly, and will keep it into consideration when I assign the 5% class participation points. You are responsible for what is covered in class whether you were there or not.

6. **Use of books and notes on tests** All tests will be close-books and will be proctored.

7. **Calculator** You should bring the calculator to class every day. You are required to bring the calculator to all tests, EXCEPT GATEWAY TESTS.

Appendix II-b

STUDENT DATA SHEET

1. Your name?

2. What would you like to be called?

3. Where are you from?

4. What is your local address?

5. What is your local telephone number?

6. What is your e-mail address? (You will need one.)

7. What is your student ID #?

8. Are you registered in this section?

9. What College are you in (LS&A, Engineering, ...)?

10. What is your year in college?

11. What is your major (or main interest, if no major)?

12. About how many students were in your high school graduating class?

13. Why are you taking this course?

14. Please list all **math c**lasses you've taken in high school and college, and describe them briefly (for example; they were challenging, they were boring, they were confusing, loved them, hated them,...).

Class	When Taken	Where Taken	Description
____________	______________	______________	____________________________
____________	______________	______________	____________________________
____________	______________	______________	____________________________
____________	______________	______________	____________________________
____________	______________	______________	____________________________
____________	______________	______________	____________________________

15. Other comments about mathematics, the University, or life in general.

What Needs to be Accomplished

Establish course goals.

Start learning and using names.

Get students acquainted with each other and you.

Gather information about the students.

Communicate how the class will be conducted.

Get students involved in a group activity.

Give a clear statement of the assignment.

Appendix III-a

Guide for Analyzing Explanations

Material and Content

____ 1. Instructor states the topic clearly at the beginning.

____ 2. The explanation is organized.

____ 3. Instructor emphasizes the main idea.

____ 4. Instructor relates material to something students already know.

____ 5. Instructor gives suitable examples.

____ 6. The material is appropriate to the level of beginning undergraduates.

____ 7. Instructor has a clear conclusion.

Technique and Delivery

____ 8. Instructor involves students (eye contact, questions, etc.).

____ 9. The instructor shows enthusiasm.

____ 10. Board work is clearly organized and readable.

____ 11. The pace is appropriate.

Comments:

Appendix III-b: Overhead

Definitions taken from Webster's Dictionary

Lecture:

1. An informative talk given before an audience, class, etc., and usually prepared beforehand.

2. A lengthy scolding.

Explain:

1. To make clear, plain, or understandable.

2. To give the meaning or interpretation of; expound.

Appendix III-c

Group and Topic List for Practice Teaching Sessions

Instructions :

1. The letter code for each of the three sessions identifies which group you will be in for that day. The key printed on the back of this page shows the meeting place and time for each group.

2. The number code for each of the three sessions identifies which topic you should prepare as a practice lesson. The list of topics, and the texts, is printed on the back of this sheet.

		Tuesday	Wednesday	Thursday
1	E. M.	A 1	F 6	K 11
2	S. P.	A 2	G 6	L 11
3	J. A.	A 3	H 6	M 11
4	P. A.	A 4	I 6	N 11
5	Q. C.	A 5	J 6	O 11
6	J. D.	B 1	F 7	O 12
7	E. F.	B 2	G 7	K 12
8	M. B.	B 3	H 7	L 12
9	G. B.	B 4	I 7	M 12
10	R. M.	B 5	J 7	N 12
11	M. K.	C 1	F 8	N 13
12	G. M.	C 2	G 8	O 13
13	M. G.	C 3	H 8	K 13
14	D. V.	C 4	I 8	L 13
15	N. S.	C 5	J 8	M 13
16	P. O.	D 1	F 9	M 14
17	J. W.	D 2	G 9	N 14
18	T. H.	D 3	H 9	O 14
19	C. W.	D 4	I 9	K 14
20	H. W.	D 5	J 9	L 14
21	J. M.	E 1	F 10	L 15
22	C. F.	E 2	G 10	M 15
23	L. J.	E 3	H 10	N 15
24	S. M.	E 4	I 10	O 15
25	L. W.	E 5	J 10	K 15

Key to Letter Codes

Group	Time	Place	Facilitator	Group	Time	Place	Facilitator
A	9:00 AM	B110	M. D.	I	10:30 AM	B113	E. G.
B	9:00 AM	B111	D. W.	J	10:30 AM	B114	W. C.
C	9:00 AM	B112	P. S.	K	11:00 AM	B110	M. D.
D	10:30 AM	B113	E. G.	L	11:00 AM	B111	D. W.
E	10:30 AM	B114	W. C.	M	11:00 AM	B112	P. S.
F	9:00 AM	B110	M. D.	N	11:00 AM	B113	E. G.
G	9:00 AM	B111	D. W.	O	11:00 AM	B114	W. C.
H	9:00 AM	B112	P. S.				

Key to Numerical Codes (List of Topics)

Note : The textbook used will be :

Hughes-Hallet : 'Calculus"

Topics for Tuesday (Practice Teaching)

1	The Number *e* and Natural Logarithms	pp. 47-51
2	Inverse Functions	pp. 35-38
3	Average and Instantaneous Velocity	pp. 94-98
4	The Second Derivative	pp. 126-129
5	Derivatives of Power and Polynomial Functions	pp. 190-196

Topics for Wednesday (Practice in Using Active Learning)

11	The Derivative	pp. 111-117
12	How Do We Measure Distance Traveled ?	pp. 150-157
13	The Tangent Line Approximation	pp. 233-236
14	Optimization	pp. 282-286
15	Interpretations of the Derivative	pp. 119-125

Topics for Thursday (Practice in Answering Questions)

6	Interpretations of the Derivative	pp. 119-126
7	Definite Integral as Area and Average	pp. 164-170
8	Fundamental Theorem of Calculus	pp. 171-178
9	The Chain Rule	pp. 212-216
10	Approximations and Local Linearity	pp. 132-138

Appendix IV-a

Guide for Analyzing a Cooperative Learning Exercise

_____ 1. Instructor provides adequate introductory information about the topic.

_____ 2. Instructor gives clear, detailed instructions about group tasks.

_____ 3. Instructor provides effective instructions for dividing the class into groups and facilitates the process as needed.

_____ 4. Instructor monitors group progress and assists groups when necessary.

_____ 5. Instructor creates an opportunity for groups to report back to the whole class or utilizes outcomes of group work in class discussion.

Overall Comments

_____ 6. The topic selected was appropriate for group work.

_____ 7. The tasks assigned to groups were challenging, interesting, and had a clear and useful purpose as group assignments.

_____ 8. Instructor utilized class time well.

_____ 9. Working in groups enhanced student understanding of the material.

Getting Students to Read the Book

Bob Megginson

1. Don't lecture as if the students have never before seen the material!

2. Don't lecture as if the students have never before seen the material!

3. Don't lecture as if the students have never before seen the material!

4. You must *really* expect them to read the book, and always act as if you expect them to read the book.

5. You must set the tone the first day of class, *saying* that they must read the book and why.

6. *You* must read the book!

7. Make each day's assignment of reading an *event*, complete with coming attractions.

8. When you start the day's activities, do a brief activity that assumes the reading of the book.

9. As a rule, *don't* do examples directly from the book, unmodified--it sends the wrong message. (But there are exceptions.)

10. When examples from the book are important and difficult, go over the *difficult* parts, only outlining the *results* of the easier parts, constantly tossing in phrases such as "as you saw in your reading,..." Better yet, get them involved in a group activity to work through the exercise themselves.

11. When they *aren't* doing the reading, try:
    (a) Brief quizzes over the reading at the beginning of the day.
    (b) Group activities based on the reading, as mentioned above.

Most Importantly--

12. Don't lecture as if your students have never before seen the material!

Appendix VI-a

List of Teams and Roles for first four weeks

Please find a seat near the number on a desk that corresponds to the number of your group. There should be three other people in your group. Please introduce yourselves and share the following information with each other:

-Where you grew up
-Why you chose the University of Michigan
-What you like to do in your free time

Group 1
1. ____________
2. ____________
3. ____________
4. ____________

Group 2
1. ____________
2. ____________
3. ____________
4. ____________

Group 3
1. ____________
2. ____________
3. ____________
4. ____________

Group 4
1. ____________
2. ____________
3. ____________
4. ____________

Group 5
1. ____________
2. ____________
3. ____________
4. ____________

Group 6
1. ____________
2. ____________
3. ____________
4. ____________

Group 7
1. ____________
2. ____________
3. ____________
4. ____________

Group 8
1. ____________
2. ____________
3. ____________
4. ____________

Roles:

Student	Week 1	Week 2	Week 3	Week 4
1.	Scribe	Clarifier	Reporter	Manager
2.	Clarifier	Reporter	Manager	Scribe
3.	Reporter	Manager	Scribe	Clarifier
4.	Manager	Scribe	Clarifier	Reporter

Homework Teams - Rules and Roles

Group homework will be a significant part of this course. Generally, there will be two assignments per week. There will not usually be many problems, but they will generally require a lot more thought and discussion than math problems you have seen before. Each member of the team has an important role. These roles are to be rotated each week so that everyone gets a chance to try each role. The roles are:

1. **Scribe-**The Scribe is responsible for writing up the single final version of the homework report to be handed in. This is the only report that will be accepted or graded. The grade will apply to each member of the team. Whenever possible, the solution of each problem should have symbolic, graphical and verbal explanations or interpretations. Diagrams and pictures should be provided whenever possible.

2. **Clarifier-**During the team meeting the Clarifier assists in clarifying for the group, ideas that other members of the team may present. The Clarifier is responsible for making sure that each member of the team understands the group's solution to the problems.

3. **Reporter-**The reporter writes a record of how the homework session(s) went, how long the team met, what difficulties or particular successes the team may have had (mathematical or otherwise). If there is disagreement about the solution of a problem, the reporter should present alternate solutions, and explain the difference of opinion. The report should list the members of the team who attended the session and their roles. **The report should be on a separate sheet of paper and be the first page of the homework report. The names of the team members and their roles should be listed on this page.**

4. **Manager-**The Manager is responsible for scheduling the team meeting (i.e., coordinating when and where the team meets) and arranging for any refreshments (pizza, etc.). It is the Manager's job to run each meeting. If one of the team members is seriously ill and cannot attend the session, the Manager assumes that individual's responsibility. If the instructor has asked the group to prepare an "overhead" for class presentation, the Manager will be responsible for that task.

The various roles should be rotated around the team. In some classes, the instructor will determine the roles, in others the teams will do that themselves. Remember, late homework will not be accepted.

Cooperative Behavior

When working in groups with other students, the goal is for all of you to cooperate in the learning of all members of the group. In other words, when you are finished with a team assignment **everyone in the group should understand and be able to explain how to solve the problems**. The ideas listed below are meant to help each group member and each group work at their full potential.

- Be prepared. Do the readings , look over the assignment and try to do the problems before the group meets. Be prepared to discuss, explain, and/ or ask questions.
- Listen carefully and with respect to each other.
- Criticize ideas, but do not criticize people.
- Everyone has the right and responsibility to contribute to the task on which the team is working. In other words, when you are finished with an assignment, everyone in the group must understand and be able to explain how to solve the problems.
- Ask for help when you need it.
- Give help when it is requested.
- Make decisions by reaching consensus, not by majority rule. Don't agree to something you don't understand.
- Do not allow one or two members of the group to dominate the discussions. This can be very damaging to successful group activity.
- From time to time the members of the team will be asked to do a peer evaluation of the group and its members (See the "Team Evaluation Form" on the next page.)

Team Evaluation Form

Your name:____________________________

Date_________________

Instructions: Please enter the names of your team members, and enter your evaluation as follows

not a strength = *0*
ok = *1*
a real strength = *2*

	A:	B:	C:
Name of Team Member	________	________	________
Attendance at group meetings	________	________	________
Has read the material and tried to work the homework problems	________	________	________
Comes on time	________	________	________
Helps keep the group going	________	________	________
Willing to listen to others	________	________	________
Puts effort into the process	________	________	________
Helps clarify problems	________	________	________
Is willing to disagree	________	________	________
Is tuned in to whether other members of the group understand the problem	________	________	________
Helps assure that everyone understands the solution.	________	________	________

Please Circle the Appropriate Response:

Being part of this group helped me better understand the material
Strongly Agree Agree Neutral Disagree Strongly Disagree

Meeting with this group was better than trying to work the problems on my own.
Strongly Agree Agree Neutral Disagree Strongly Disagree

Meeting with this group was a good experience.
Strongly Agree Agree Neutral Disagree Strongly Disagree

What suggestions would you make to improve your group or group experience? *(Please use other side)*

Checklist for Homework

1. Did you include a Reporter's page?

2. Is the homework neatly presented (no fringes, stapled, legible)?

3. Do you answer the question asked? Reread the question to be sure.

4. Have you used the rule of three? Where possible have you included algebra, numerical examples or tables, and a graph (that is clear and labeled) to support your answer?

5. Are your answers readable? Have you used complete sentences to describe the procedure by which the answer is derived? Did you justify your answer? Would someone who had not read the question understand what was going on? (Don't rewrite the question). Read your answer back. Is it clear?

What is Important in Forming Teams

- Quickly get students into their homework groups

- Give students an activity to get acquainted with each other.

- Give teams a chance to work together in class.

- Discuss how team work relates to course goals

- Outline students' responsibilities to their team

Appendix VII-a

Group Observation Worksheet

1. Members know and use each other's names.

2. The group agrees on a plan of attack.

3. Everyone seems to be working on the problem.

4. The group remains focused on the problem.

5. Everyone has the opportunity to express his/her own views.

6. Members listen to each other.

7. Members ask for help when needed.

8. Members give help when asked.

9. The group has a mechanism for assuring that everyone understands the problem.

Other observations:

15. The Montgolfier brothers (Joseph and Etienne) were 18th Century pioneers in the field of hot-air ballooning. Had they had the appropriate instruments, they might have left us a record of one of their early experiments, like that shown in Figure 3.29. The graph is of their vertical velocity, v, with upwards as positive.

    (a) Over what intervals was the acceleration positive? Negative? Zero?

    (b) What was the greatest altitude achieved, and at what time?

    (c) At what time was the magnitude of the acceleration greatest? What was the direction of their acceleration at that time?

    (d) What might have happened during this flight to explain the answer to (c)?

    (e) This particular flight ended on top of a hill. How do you know that it did, and what was the height of the hill above the starting point?

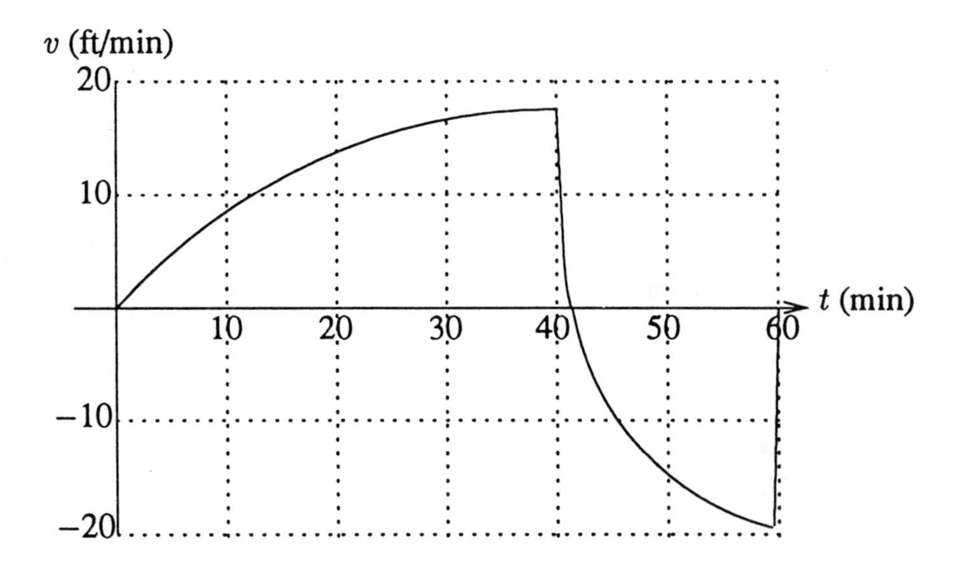

Figure 3.29: Flying with the Montgolfiers

Taken from:
Calculus, by Deborah Hughes-Hallett, Andrew M. Gleason, et al. Published by John Wiley & Sons, Inc.

13. Let $f(v)$ be the energy consumption of a flying bird, measured in joules per second (a joule is a unit of energy), as a function of its speed v (in meters/sec). See Figure 5.61.

(a) Explain the shape of this graph (in terms of the way birds fly).

Now let $a(v)$ be the energy consumption of the same bird, measured in joules *per meter*.

(b) What is the relationship between $f(v)$ and $a(v)$?

(c) Where is $a(v)$ minimal?

(d) Should the bird try to minimize $f(v)$ or $a(v)$ when it is flying? Why?

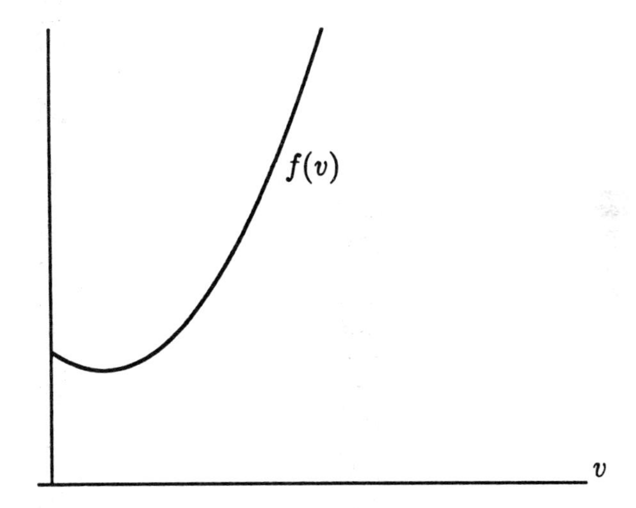

Figure 5.61: Energy Consumption of a Flying Bird

Taken from:
Calculus, by Deborah Hughes-Hallett, Andrew M. Gleason, et al. Published by John Wiley & Sons, Inc.

Class Management

Organize the information flow.

1. **State clearly all policies and expectations about class procedures such as tests, homework, and student behavior.**

2. **Announce homework far enough in advance, collect it regularly, grade it efficiently, and return it promptly.**

3. **Establish fair grading policies and convey them unambiguously to the students.**

Plan carefully and use the time well.

1. Stay with the syllabus.
2. Schedule well in advance for quizzes, review, student presentations, etc.
3. Organize the class period. Include an overview at the beginning of class and close with a smooth summary.
4. Balance lecturing and cooperative group work.
5. Readjust plans when you are suddenly faced with too much time or too little time.

Appendix VIII-c

MATH 115 -- FALL, 1995

SAMPLE INTERACTIVE LESSON PLANS

This organization is appropriate for classes which meet for 50 minutes on a M,T,W, F schedule with team homework due twice a week and quizzes on Fridays. If your class operates differently, you will have to make some adjustments. The text references are for *Calculus*; Hughes-Hallett, Gleason, et al but the interactive design of class time is appropriate with any challanging problems.

Tues. Sept. 5

10-25 (minutes) Class officially starts at 10 past the hour--try to arrive a few minutes early to write the assignment on the board and greet students. Give out the course syllabus and a sheet describing your class; your name, office location, office hours, telephone, e-mail, grading, homework routine, etc.
Have them fill out the student data forms, and turn them in when they finish. These forms will help you learn names and assign teams.

25-30 Introduce yourself and tell them about the course. Make sure that you state that reading the text, writing clear explanations, and cooperative learning are fundamental aspects of the course.

30-45 Ask the students to form groups of four with their neighbors and introduce themselves to each other. Do section 1.1 #1 as a group. Let them take several minutes to read and think about the problem on their own before talking about it with each other. Tell them before they start that you will be calling on someone fom one of the groups to read the group's story for the missing graph to the whole class.

45-60 Have them work on section 1.1 #6 as a group. This problem seems strange at first because they tend to confuse the plane's path with the graph. It helps to have them think about exactly what is being graphed and label the axes. If you have time have a few groups put their graphs on the board..

Assignment
1) Get text, calculator
2) Read the textbook's Preface for Students, read section 1.1
3) Do individual problems 1.1 #3, 5, 7, 11, 13
4) Prliminary team problem, on handout due Friday

Wed. Sept. 6 --- last night's reading, section 1.1

10-15 Ask an easy question or two about the reading to reinforce the importance of doing it. Call on students by name to answer.

15-30 Give a brief summarizing lecture on section 1.1 spending a few minutes on proportionality. The book doesn't treat it too heavily, and it will be used throughout the course. Maybe you can get the students to come up with some examples. *What would a table look like? What would a graph look like?* Comment that the rest of Chapter 1 continues discussing certain types of functions in detail, starting with linear functions (they've just seen an example; proportional functions).

30-55 Get them into their homework teams. Give them a list or somehow tell them which teams they're in. Have them move so they are with their group members and tell them to introduce themselves to each other and exchange local addresses and phone numbers. Make sure they understand the different roles of each team member and what it means to be a coopeerative member of a group. It might be a good idea to have each team assign its roles right away and have the manager schedule the first meeting.
Now have them work on 1.2 #15 treating it as if it were a team problem due on Friday. Help them come up with an accurate diagram of the situation and then label the axes.They'll often see it more clearly if they make a table before graphing anything (rule of four). They should come up with a single solution, nicely written up, and understood by each team member. If there's time have a clarifier explain it at the board.

55-60 Announce a quiz for Fri. covering the 1.1 and 1.2 readings and the homework problems from 1.1 (they can check their solutions to these in their homework teams).

Assignment
1) Read 1.2
2) Do individual problems 1.2 #1,3,6,10,13,20
3) Start on the team problems due on Tues. Sept.12

Fri. Sept. 8 --- last night's reading, section 1.2

10-25 Ten minute quiz allowing them to use calculators whenever they want. Remind them that in a cramped classroom they should concentrate strictly on their own papers.

25-30 Ask them to do 1.2 #8 quickly without a calculator. *Through the origin? Positive slope? Negative slope? Positive y-intercept? Negative y-intercept?*

30-45 Have them get in their homework teams. Return their Practice Team Problems and have them exchange papers with another team. Lay out the criteria and point scheme that you plan to use when grading their team problems and have them practice by grading the other team's problem on a separate sheet of paper. Then have the groups explain their grading to each other. You may want to collect both their solutions and their corrections; not to assign a score, just to have a look.

40- 00 Preview 1.3 a bit by asking students what they think an exponential function is. Try using the *Ideas for the class*"on p.10 of the green Instructor's Manual to help them organize the main ideas.

Assignment
1) Read 1.3 2) Do individual problems 1.3 #3,4,5,7,9-11 3) Team problems due on Tues. Sept. 12

Mon. Sept. 11 --- last night's reading, section 1.3

10-15 Return their homework problem and quizzes and comment on the results.

25-40 This is probably a good time to do some lecturing covering the shape of exponential graphs, asymptotes, and concavity. Students often ask which exponential formula to use so it's a good idea to explain why they're just different forms of the same formula. Go over how to find a specific exponential formula from a table or from a graph. They may also need help working with the exponents (as in ex.2) and in taking roots on the calculator.

40-55 Explain the idea of radioactive decay and half-life, but tell them that you'll get back to carbon dating in a later section when there are some problems about it. You may want to present the ideas by going through problems with them. Start by leading them through #17, doing it as a class. Then continue with #19.

55-00 Remind them that the Math Lab will point them in the right direction if they need help.

Assignment
1) Reread 1.3 2) Do individual problems 1.3 #12, 13, 15, 19 3) Finish team problems due Tues. Sept 12

Tues. Sept. 12 --- last night's reading, section 1.3

10-20 Collect the team homework at the start of class. Solicit questions on any homework.

20-35 Have them work in small groups on #20, then call on a group to present the solution. After they've finished you might mention that knowing just two data points can lead to wildly varying predictions about the future behavior of a function.

20-25 To recap the difference between linear functions and exponential functions, call on different students for the following: How can you tell the difference if you are given: a table? a graph? a formula? a description? Ask if there are questions at all on 1.3. If they don't ask any, write y = 3(1- (.06)^x) and ask "What happens when x gets big? When x gets close to zero? What's the thing look like?"

35-55 Read the green instructor's manual to get a feel for 1.4 and exercises. Problem #20 in 1.4 is good for them to work on now; it's hard but non-threatening.

Assignment:
1) Read 1.4 2) Do individual problems 1.4 #7,11,13,17 3) Work on team problems due Fri. Sept. 15

Wed. Sept.13 --- last night's reading, 1.4

10-20 Return previous homework and discuss any interesting answers. Remind them that learning to write well is one of the course goals, and that you're pleased with their first assignments.

20-25 Talk about the procedure of modeling. Assume you know or have some idea of what the function looks like eg. $y=k(a)^t$. Then you use the information to rewrite the formula for your specific problem. Only then can you use the new formula (say $y=5(3)^t$) to make predictions.

25-35 Have them do #15 pg.34 in groups, having them think of it as modeling with a power function. If one of the groups finishes early, have them figure out the top speed that the car can go and still stop in 200 ft.

35-40 When they've finished #15 have each group work together to write up the problem as if it were a team homework problem - don't tell them what you're going to do with the solutions until they've finished.

40-50 Pass papers between groups and have each group grade another group's paper. Use your own homework number grades for a problem eg. 10/10 or 3/5, etc.

50-00 Discuss what you would expect a good solution to be: paraphrase of the problem, rule of 4, clearly labeled graph, etc. Pass out a sample solution showing what elements are necessary for a perfect score. Remind them to use appropriate variables, not just x's and y's every time.

Assignment
1) Read 1.5 2) Do individual problems 1.5 #1,2,3,6,9,12,13 3) Team problems due Fri.. Sept.15

Fri. Sept. 15 --- last night's reading 1.5

10-15 Collect the homework that's due and hand back the corrected homework.

15-30 Quiz.

30-35 Recap inverse functions. Get them to tell you that finding f(5) means finding the y value that comes from using x=5, but finding $f^{-1}(20)$ asks you to find what x value gives you 20. They have trouble with the notation (no wonder, it's absurd). They may confuse the inverse with either the negative -f(x) or the reciprocal $(f(x))^{-1}$.

35-45 Quickly get them into groups with their neighbors and answer the question: "Why is the statement in the top blue box on pg.38 true?" Choose a student at random to explain it to the class. If anyone gets lost or confused in this reading exercise, take the time to straighten it out.

45-55 Have them graph 10^x on [-3,3]x[-3,3]. What would an graph inverse look like? What points are on it?

55-00 Remind them (and yourself) that they've gotten through the first TWO WEEKS intact!

Assignment
1) Read 1.6 2) Do individual problems 1.6 #5,9,13,15,20,24, 27 3) Team problems due Tues. Sept. 19

Mon. Sept. 18 --- last night's reading 1.6

10-12 Hand back homework. Tell them that on Tues. you'll show them some good solutions done by the class.

12-20 Write the log rules (both base 10 and ln side by side) on the board for reference. Take homework questions.

20-25 Hand out the log worksheets (one is in your mailbox) and begin by doing #1 and #2 together as a class.

25-45 Have them finish their worksheets. Early finishers can help others.

45-55 To reinforce all this visually, have them pair off to do #23 and #24 on pg.52. Then have them go over ex.2 pg.48 thoroughly so that both partners understand.

55-00 Talk about e as a "plain old number". Without the calculator figure out e^0 and e^1. Then have them calculate e^{-1}, $e^{1/2}$, $e^{-1/2}$.

Assignment
1) Read 1.7 2) Do individual problems 1.7 #2-22 (as needed) ,27,31,33 3) Team problems due Tues. Sept. 19

Tues. Sept. 19 --- last night's reading 1.7

10-20 Collect the homework and distribute a few photocopied solutions of good answers to the previous homework team. They'll be impressed by some of the excellent work. It works best if you use a "collection" of solutions with no more than one problem from a given team. That way they see that many of their classmates are doing good work. This is a good time to remind them that all team members should keep copies of the group's homework - not just the scribe._

20-35 Do #31 pg.52 in groups. Then go over it by calculating some values (f(1),f(2),f(-1)) without graphing. Fool with the ideas like f(large)=? , f(close to zero)=?, f(large negative)=? etc.

35-37 You can preview a team homework problem by telling them the story of the Art Forger whose specialty was Vermeer. Just to spite the art historians who still doubted his forgeries, while he was serving his jail sentence, he requested some old paint and brushes and forged another!

37-55 Without using any numbers or formulas, just define in English and explain the terms: Principal, Balance, Annual compounding, Quarterly compounding, Daily compounding, Countinuous compounding, Yield, Effective annual yield, Annual percentage rate (APR), Nominal rate. Help them get the idea of what's going on here without their being blinded by the vocabulary.

55-00 Tell them to gather up their unanswered questions about logarithms and expontials. You will answer them on Wed.

Asssignment
1) Read 1.8 2) Do individual problems 1.8 #1,3,7 p.87#15,16 3) Work on team problems due Fri. Sept.22

Wed. Sept. 20 --- last night's reading 1.8

10--25 Collect the team homework. Answer any questions.

25--35 They usually need a short but careful review of functional notation. Try writing the functions in 1.9 #2 as

$$f(\;) = 2(\;) + 3 \qquad\qquad g(\;) = \log(\;)$$

then have them form things like: f (5), g (10), f (t), g (t), f (x+1), f (x) + 1, f (x + h), f (x + h)--f (x), f (g (x)), g (f (x)),and especially [f (x+h)--f (x)] /h.

35--00 To work on all this graphically, tell them that they should memorize the graphs on the handout sheet (it's in your mailbox). Ask them to graph y=x^2 on the window [--5,5] x [--5,10]. Have them experiment with the affect of multiplying a function by a constant by graphing y=2(x^2) on the same axes. Follow up with y=(x^2)+1 and y=(x^2)--1 and y=(x--1)^2 and y=(x+1)^2. Then have them form groups to work on 1.9 #16 asking them to describe to each other exactly what has happened to the graph each time.

50--00 Have them read #22, an upcoming team problem. They're often baffled by this problem. If you hint, "the rule of three", they may think of making a table which will start them on the right track. Whenever you go through this problem, you'll have an opportunity to reinforce the idea that 1/ (small) = large, and 1/(smaller) = larger.

Assignment
1) Read 1.9 2) Do individual problems 1.9 #1,5,6,7,8,23,29,30 3) Finish team problems due for Fri.

Fri. Sept. 22 --- last night's reading 1.9

10--15 Collect and hand back homework.

15--30 Quiz

30--55 Ask them to compare answers to the homework problems 1.9 # 29,30 making sure that they can articulate the patterns on tables with the patterns on graphs. When they are done, have them do # 17 in groups and explain their answers at the board. To emphasize the idea of teamwork, don't let the person who writes the solution on the board do the explaining.

55--00 Preview the next section by recalling that so far they've studied several families of functions in detail (have them list them; linear, exponential, logarithmic, power). Each family had its own particular properties with a characteristic table and graph. The next functions they will look at are the trig functions. They will be looking at these as functions, not from the point of view of triangles, in particular, they should try to get used to radian measure instead of degrees.

Assignment
1) Read 1.10
2) Do individual problems 1.10 #1-18,20,21,25,33-40,41
3) Work on team problems due Tues. Sept. 26. Tell them that next Tues. you **plan to call on a team** at random to present one of the team problems so each student should be prepared if called on.

Mon. Sept. 25 --- last night's reading 1.10

10--20 Ask them a few brief questions on the reading; what's amplitude?, what's the general equation for the sine? what do the graphs look like?

20--55 Then tell them that they're going to do three problems displaying the "rule of three".
42 ....words to formula # 43..........graph to formula # 47..........table to formula
Have them count off so they are in groups of three, move together and introduce themselves. When they've all finished the first one, summarize it yourself. Then go on to the next.

55--00 Spend a few minutes asking them how to go in the other direction; from formula to words, formula to graph, formula to table.

Assignment
1) Read 1.11 paying particular attention to which features of a function's formula cause zeros and asymptotes. Could the graphs of polynomials have asymptotes? Why, or why not?
2) Do individual problems 1.11 #1,5,7,9,13,17,21,25 3) Finish the team problems and prepare to present.

Tues. Sept. 26 --- last night's reading 1.11

10--20 Collect the homework and select a correct solution to 1.9 #22. Call on a team member, other than the scribe, to present the solution. Have the student put the original graph high on the board and the reciprocal graph directly underneath. Make sure the student asks for questions before sitting down. See that the class understands what causes the reciprocal graph to increase so quickly .

20--40 Have them tell you how to graph y = x--2 (without calculululators), and then you put the graph high on the board. Then ask them to imagine the graph of y = 1/(x--2). Now have them make up a function which has a zero at x=2 and an asymptote at x=1. Graph it on zoom-2; does it work? What about (x--2)/(x--2)?? Now get them to review with you how the graphs of polynomials look. You might try some suggestions from *Ideas for class p.18.*

40--55 Problem #14 is important for them to understand. Start by sketching the actual up-down trajectory and saying that at t=0, f(t)=0. Go through the problem slowly and both algebraically and graphically following the outline used in example 3. Each time you associate a specific height with a specific instant, mark it on the trajectory. After you finish, tell them that as part of tomarrow's homework you want them to read section 1.11 example 3 and then come to class ready to discuss #15.

55--00 Wrap up the chapter by telling them that they have now studied in detail a library of functions; linear, exponential, logarithmic, power, trigonometric, polynomial and rational . They can work with them symbolically, graphically, numerically, or from a description. Now they are ready to begin the study of calculus -- starting with the idea of the derivative.

How Doug Marks Papers

by Doug Shaw

The following is a description of how I convert a stack of raw homework to a stack of corrected, graded papers. I've tried many different methods, and this one (for me) gets the papers marked fairly and well, in the least amount of time.

I first "correct" each problem as completely as I can. Whether I correct one assignment at a time, or all the #1s, then all the #2s etc. depends upon the assignment. When they are all corrected, I go through and assign each problem a grade from 1-5, then total them up. I refuse to give half-points, because that effectively changes the scale from 1-10. Keeping it from 1-5 makes things expedient and (I believe) more just. Sure, grading on a course scale requires using judgment. We are professionals; that's what we're being paid for.

5: Not necessarily "perfect." Some write-up problems are okay. A couple of mathematical equivalents of "typos" are okay. An intelligent, thoughtful, but incorrect approach may get this score, too. In general, a (possibly flawed) work of art.

3: Serious write-up problems (although some effort shows) with good mathematics. Or a seriously flawed approach written up well. Or a combination thereof. A three problem has the feeling of, "Look, let's just write up what we have so far and move on."

1: This means that some minimal effort was made, but they don't have the concept. If someone just writes down the right number, but shows minimal work, I will give this grade. (I would give a zero if they show <u>no</u> work, but this has never happened to me.) If someone writes absolute gibberish, but puts effort into the write up, I will give this grade. To me, this grade means they didn't leave it blank, but they definitely didn't get the concept.

I reserve the 4 for when I can't decide between a three and a five, and the 2 similarly. That way I don't need to take too much time assigning numbers. I think, "Five, three, or one?" and I have two "in-between" grades if I can't quickly decide. I find that I take some time on the correcting, but the actual assigning of grades usually goes fast, and I'm usually pleased with the result.

A Few Suggestions on Grading

Bob Megginson

1. Students' complaints about "nitpicking" usually are caused by the instructor's littering the problem with - 1/2's and - 1's for individual errors. It is usually far better to read the entire exercise (or part of the exercise, if it is divided into parts (a), (b), and so forth), correcting errors as you go, and then assign one number as the grade for the entire unit consisting of the exercise or part. In short, you should *correct* all errors, no matter how small, but the smallest unit to which a *number*, either positive or negative, should be attached is an exercise or part of an exercise. (Now see the next suggestion about negative numbers in grading.)

2. Students often seem to react better when you give them points rather than take them away. If an exercise is worth 8 points and the student misses two of those points, it seems to be better to write 6 beside the exercise rather than - 2. Also, using a "give" rather than a "take away" system can lead to better grading, since the grader is more likely to think of what the effort is worth overall rather than which individual errors require deductions.

3. There is not universal agreement on the value of this, but I like to grade one problem at a time rather than one paper at a time. That is, I will grade everyone's exercise 1, then everyone's exercise 2, and so forth. For me, it fosters uniformity in grading and makes it less likely that I will be abnormally cruel to a student for a mistake in exercise 3 after the student has totally bombed exercises 1 and 2.

4. Just as you would not shout disapproval at students in class, you should not shout disapproval on paper. In comments that could be construed to be negative, you should avoid such emphatic devices as underling, capitalizing, and exclamation marks, since these are just written forms of shouting that can hurt students' feelings. Put yourself in the student's shoes when reading the following two remarks:

   This problem cannot be done this way since the function is not linear.

   This problem **CANNOT** be done this way since the function is **NOT LINEAR!!!**

   The first is a statement of fact, and the student should not take it amiss. The second says, essentially, that the grader considers the approach (and the student taking the approach) to be particularly misguided. It may be true that the grader does consider the approach to be so, but nothing is to be gained by saying that. Pointing out the error is sufficient. So what about the following compromise?

   This problem cannot be done this way since the function is obviously not linear.

   This is perhaps not quite so bad, but what is really added by the word "obviously"? The grader is effectively saying that the student is somehow deficient for not noticing something that should be "obvious." *In general, avoid the use of emphatic devices such as underlining, all capitals, exclamation points, and emphatic adjectives when correcting errors.* Of course, it is perfectly fine to use those devices in *positive* comments!

5. The preceding item leads to the question of what to do when the student really is bluffing and what is written is nonsense. While the temptation is strong to retaliate, the point is adequately conveyed by a zero grade and a neutral comment such as "This is incorrect--see the examples given in class."

Why Get Students to Write?
Dale Winters

Writing and revision of writing are learning experiences that lead to conceptual understanding.

What do students get out of writing?

• *Writing an explanation of their work:*

1. *requires that students organize and order their work*
2. *helps them identify specific points, concepts and work where understanding is poor*
3. *helps them produce clear, well articulated answers*
4. *helps them organize and structure knowledge in their minds*
5. *encourages students to present homework solutions in a systematic, logical order.*

• *Using a verbal or written form of expression to describe abstract, technical concepts deepens and broadens students' understanding by giving them new ways to express concepts and ideas.*

• *Expressing the key points as written sentences improves students' ability to remember them.*

• *A clearly explained, well articulated and correct solution to a problem provides a permanent record of the group's thoughts on solving the problem. This is far more valuable to the students when they study for an exam than a list of algebraic manipulations, whose point (and probably even the meaning of the symbols themselves) is easily forgotten.*

What do instructors get out of having students write?

• *Writing and review deepens students' understanding, which ultimately makes our job easier.*

• *Students hand in more logical, systematic, clearly thought out, and articulate homework solutions, which are easier to grade.*

• *Students gain a sense of pride in their work, and with this pride can come a deeper interest in the class.*

• *Requiring students to write explanations of their work serves as a diagnostic tool for the instructor. Written solutions can reveal gaps and flaws in students' understanding:*

1. *omissions in the explanations*
2. *faulty reasoning and assumptions are recorded*
3. *evasions and inconsistencies in thinking are evident.*

• *Requiring students to explain the origins of, and reasoning behind, the mathematical models they devise to solve problems clarifies the process of mathematical modeling and the links between mathematics and the real world.*

Appendix X

15. According to the April 1991 issue of *Car and Driver,* an Alfa Romeo going at 70 mph requires 177 feed to stop. Assuming that the stopping distance is proportional to the square of velocity, find the stopping distances required by an Alfa Romeo going at 35 mph and at 140 mph (its top speed).

Taken from:

Calculus, by Deborah Hughes-Hallett, Andrew M. Gleason, et al.
Published by John Wiley & Sons, Inc.

15. According to the April 1991 issue of *Car and Driver*, an Alfa Romeo going at 70 mph requires 177 feet to stop. Assuming that the stopping distance is proportional to the square of velocity, find the stopping distances required by an Alfa Romeo going at 35 mph and at 140 mph(its top speed).

Taken from: *Calculus*, by Deborah Hughes-Hallett, Andrew M. Gleason, et al. Published by John Wiley & Sons, Inc.

Section 1.4 Problem #15

Given:
70 MPH → 177 ft to stop - this was supplied to us in the problem,

$y = f(x) = Kx^P$ (K and P are our constants)

K = variable + stopping distance is proportional to the velocity^2

$$\frac{70^2}{177} = K$$

$$Y = KX^7$$

$$Y = \frac{70^2}{177} \cdot MPH^2$$

@ 35 mph

$$Y = \frac{70^2}{177}) \cdot 35^2$$

$$Y = 44.25 \text{ ft.}$$

@ 140 MPH

$$Y = (\frac{70^2}{177} \cdot 140^2$$

$$Y = 708 \text{ ft.}$$

15. According to the April 1991 issue of *Car and Driver*, an Alfa Romeo going at 70 mph requires 177 feet to stop. Assuming that the stopping distance is proportional to the square of velocity, find the stopping distances required by an Alfa Romeo going at 35 mph and at 140 mph(its top speed).

Taken from: *Calculus*, by Deborah Hughes-Hallett, Andrew M. Gleason, et al. Published by John Wiley & Sons, Inc.

Section 1.4 Problem #15

Given:

70 MPH → 177 ft to stop - this was supplied to us in the problem, we have two constants and our variable is going to be an exponent; So we have a power function, and it should be fit into the equation

$y = f(x) = Kx^P$ (K and P are our constants)

K = variable + stopping distance is proportional to the velocity²

$\frac{70^2}{177} = K$

$Y = KX^7$

$Y = \frac{70^2}{177} \cdot MPH^2$

@ 35 MPH

$Y = \frac{70^2}{177}) \cdot 35^2$

$y = 44.25$ ft.

@ 140 MPH

$Y = (\frac{70^2}{177} \cdot 140^2$

$Y = 708$ ft.

The rate at which the stopping distance increases is an increasing power function.